Systems & Control: Foundations & Applications

Margreet Kuijper

First-order Representations of Linear Systems

Birkhäuser
Boston • Basel • Berlin

[redacted]

Margreet Kuijper
Department of Mathematics
University of Gröningen
Gröningen, The Netherlands

Library of Congress Cataloging In-Publication Data

Kuijper, Margreet, 1961-
First-order representations of linear systems / Margreet Kuijper.
p. cm. -- (Systems & control)
Originally presented as the author's thesis (doctoral)--Centre of Mathematics and Computer Science, Amsterdam.
Includes bibliographical references and index.
ISBN 0-8176-3754-0 (Boston : acid-free). -- ISBN 3-7643-3754-0 (Basel : acid-free)
1. Control theory. 2. Linear time invariant systems. I. Title II. Series
QA402.3.K753 1994 94-16937
003'.8--dc20 CIP

Printed on acid-free paper

Birkhäuser

ISBN 0-8176-3754-0
ISBN 3-7643-3754-0
Typeset by the Author in TEX.
Printed and bound by Quinn-Woodbine, Woodbine, NJ.
Printed in the U.S.A.

9 8 7 6 5 4 3 2 1

[redacted]

Contents

Preface

This book is about the theory of system representations. The systems that are considered are linear, time-invariant, deterministic and finite-dimensional. The observation that some representations are more suitable for handling a particular problem than others motivates the study of representations. In modeling a system, a representation often arises naturally from certain laws that underlie the system. In its most general form the representation then consists of dynamical equations for the system components and of constraint equations reflecting the connection between these components. Depending on the particular problem that is to be investigated, it will sometimes be useful to rewrite the equations, that is, to transform the representation. For this reason it is of special importance to derive transformations that enable one to switch from one representation to another.

A new approach of the past decade has been the so-called "behavioral approach" introduced by Willems. One of the main features of the behavioral approach is that it is well suited for modeling the interconnection of systems. It is for this reason that the behavioral approach is a natural choice in the context of modeling. In this book we adopt the behavioral approach: we define a system as a "behavior", that is, a set of trajectories whose mathematical representation by means of differential or difference equations is nonunique. An aspect of this approach that is important in the context of representation theory is the fact that a natural type of equivalence arises.

The book focuses on four types of first-order representations that commonly arise in the process of modeling. All representations have a "generalized state space". A connection is made with the existing literature on generalized state space representations (also called "implicit systems"). In the literature these have been used primarily for systems with a nonproper transfer function. In this book we also pay attention to situations in which no transfer function exists at all. This is simply due to the fact that the existence of a transfer function has no specific meaning in the behavioral approach.

A detailed overview of the contents of this book is given in Chapter 1. The chapter further provides an introduction to the subject and motivates the approach. Chapters 2 and 3 contain basic material from linear algebra and linear systems theory. The material presented in these chapters forms the basis for Chapter 4, which deals with minimality and transformation groups, Chapter 5, on realization methods, and Chapter 6, in which structural invariants are studied. Examples will be presented throughout the book. Conclusions are given in a final chapter.

Acknowledgements

This book is based on my PhD dissertation, which I wrote at the CWI (Centre of Mathematics and Computer Science), Amsterdam, The Netherlands. I wish to express my gratitude to my former supervisor Hans Schumacher. I thank him for his many valuable comments. His numerous suggestions and ideas have substantially contributed to this book. The book was completed at the University of Groningen. I thank Jan C. Willems for his stimulating and useful advices during this last phase.

Groningen, March 1994

Margreet Kuijper

First-order Representations of Linear Systems

1

Introduction

In system theory a dynamical system is often considered as a set of differential or difference equations; these equations describe the relations between the system variables. In this book we choose a different approach that has been introduced by J.C. Willems in [64]. In this approach, the variables that are involved in the interaction of the system with its environment are called "external variables". All external variables are treated on an equal footing; in particular no decomposition in inputs and outputs is made a priori. The relationships between the external variables reflect the system laws and give rise to a family of time trajectories. A dynamical system is then defined in terms of this family of trajectories of the external variables, which is called the "behavior" of the system. Equations that describe the relations between the external variables constitute a representation of the system. In this way a distinction is made between "system" and "representation" and a natural concept of equivalence turns up: representations are equivalent if the corresponding behaviors are the same.

There is a simple reason for not making an a priori decomposition in inputs and outputs: in practice it is often not clear which of the variables should be regarded as inputs and which as outputs. As an illustration, let us consider the following example that will play a role throughout this book.

Example 1.1 Consider the electrical circuit of Figure 1.1 consisting of a capacitor and two resistors. From Kirchhoff's laws it follows that the variables $\boldsymbol{V}$ and $\boldsymbol{I}$ are related as

$$cr_2\dot{\boldsymbol{V}} + \boldsymbol{V} = cr_1r_2\dot{\boldsymbol{I}} + (r_1 + r_2)\boldsymbol{I}$$

or, in a different notation, as

$$\begin{bmatrix} cr_2\frac{d}{dt} + 1 & \quad - cr_1r_2\frac{d}{dt} - r_1 - r_2 \end{bmatrix} \begin{bmatrix} \boldsymbol{V} \\ \boldsymbol{I} \end{bmatrix} = 0. \tag{1.1}$$

In this case, it is not clear which variable is the output and which variable is the input: we can just as well choose $\boldsymbol{y} = \boldsymbol{V}$ and $\boldsymbol{u} = \boldsymbol{I}$ as $\boldsymbol{y} = \boldsymbol{I}$ and $\boldsymbol{u} = \boldsymbol{V}$.

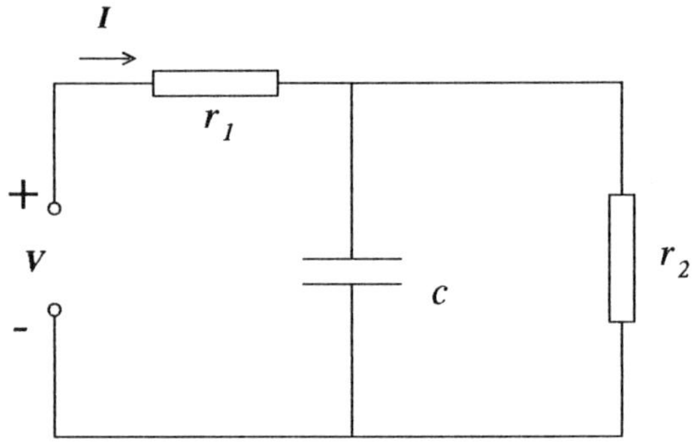

Figure 1.1: Electrical circuit I

In the behavioral approach we do not make an a priori input/output choice. In this example we would work instead with the composite vector

$$w = \begin{bmatrix} V \\ I \end{bmatrix}$$

and write (1.1) as

$$R(\frac{d}{dt})w = 0$$

where $R(s) = [cr_2s + 1 \quad - cr_1r_2s - r_1 - r_2]$. ◇

Other examples in this context are systems that consist of an interconnection of several subsystems. Such an interconnection usually induces constraints, so variables that are inputs or outputs in a subsystem can no longer be labeled as such in the interconnected system. As an example, one might think of a closed-loop system which is an interconnection of a plant and a controller. In the closed-loop system, outputs (inputs) of the plant are inputs (outputs) of the controller. Consequently, the status of the variables in the closed-loop system is not clear (unless certain constraints are imposed which enables one to label them as outputs; see [54]). The reader is referred to [68, 69, 36] for more details on the concept of feedback control from a behavioral point of view. Let us next consider the following example, which presents a situation in mechanics. The example illustrates that the behavioral approach is well suited for modeling interconnections of systems. We will come back to this example in more detail in Chapter 4 and Chapter 5.

Example 1.2 Consider two mass-spring-damper systems whose equilibria coincide; see Figure 1.2. Here M_1 and M_2 are the two masses, c_1 and c_2

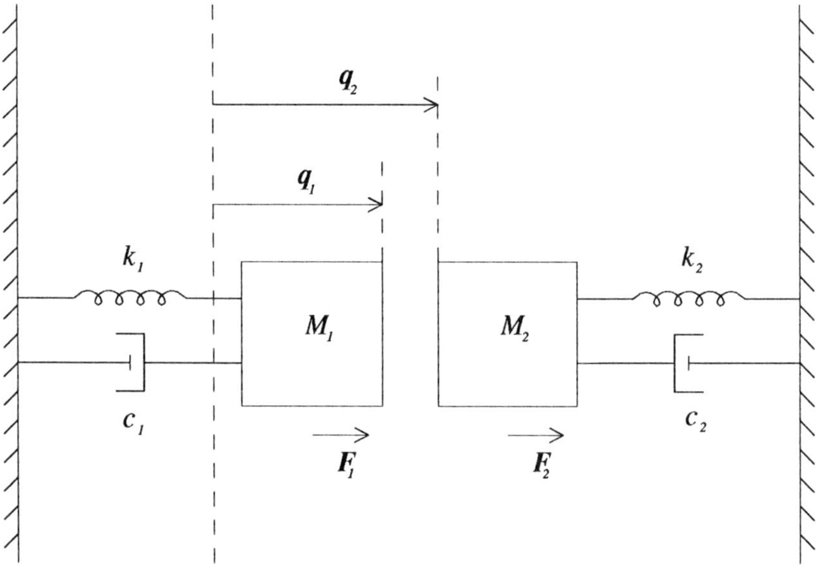

Figure 1.2: Mass-spring-damper systems before interconnection

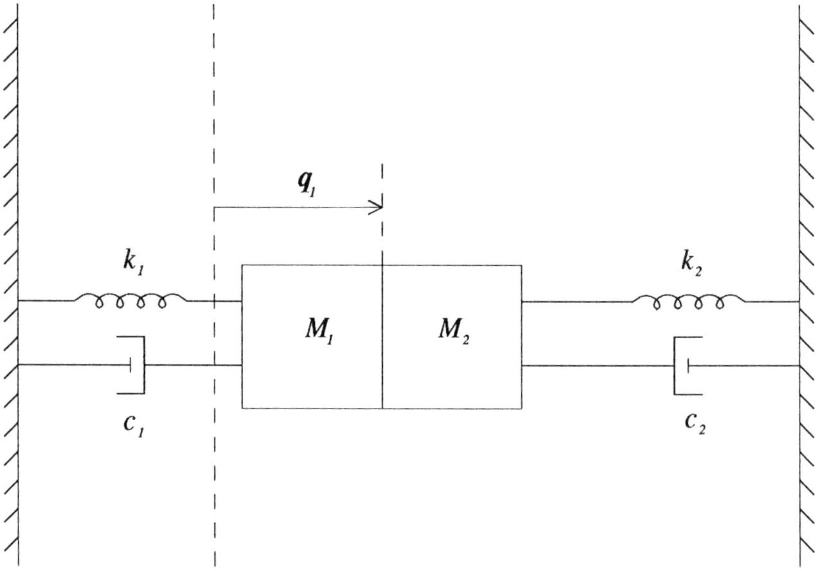

Figure 1.3: Mass-spring-damper systems after interconnection

are the spring constants, and k_1 and k_2 are the damping coefficients. A representation of the system is given by

$$\begin{bmatrix} M_1\frac{d^2}{dt^2}+c_1\frac{d}{dt}+k_1 & 0 & -1 & 0 \\ 0 & M_2\frac{d^2}{dt^2}+c_2\frac{d}{dt}+k_2 & 0 & -1 \end{bmatrix}\begin{bmatrix} q_1 \\ q_2 \\ F_1 \\ F_2 \end{bmatrix}=\begin{bmatrix} 0 \\ 0 \end{bmatrix}.$$

The external forces F_1 and F_2 clearly act as inputs. After linking the two systems together as in Figure 1.3, we get the representation

$$\begin{bmatrix} M_1\frac{d^2}{dt^2}+c_1\frac{d}{dt}+k_1 & 0 & -1 & 0 \\ 0 & M_2\frac{d^2}{dt^2}+c_2\frac{d}{dt}+k_2 & 0 & -1 \\ 0 & 0 & 1 & 1 \\ 1 & -1 & 0 & 0 \end{bmatrix}\begin{bmatrix} q_1 \\ q_2 \\ F_1 \\ F_2 \end{bmatrix}=\begin{bmatrix} 0 \\ 0 \\ 0 \\ 0 \end{bmatrix}.$$

In this system the time trajectories of the external forces are no longer independent: for example, when $M_1 = M_2$, $c_1 = c_2$ and $k_1 = k_2$, we have $F_1 = F_2 = 0$, which clearly shows that the external forces can not be labeled as inputs anymore. Note that the above illustrates that the behavioral approach allows for the modeling of interconnection as a specific type of control. The reader is referred to [68, 69, 36] for more details. ◇

The work described in this book builds on research presented in [29]–[35]. The main part of the book is concentrated upon several types of first-order representations of linear dynamical systems. Among these are representations that reflect the above set-up in the sense that no a priori decomposition of the external variable vector w is required in order to write down the representation. We distinguish two different types of representations with this property: the representation

$$\sigma G\xi = F\xi \tag{1.2}$$

$$w = H\xi \tag{1.3}$$

and the representation

$$\sigma K\xi = L\xi + Mw. \tag{1.4}$$

Here σ denotes shift/differentiation. The dynamical equation (1.2) of the first representation is determined by the matrix pencil $sG - F$; we will call the representation (1.2–1.3) a "pencil (P) representation". In fact, the P representation is a more general form of a representation that was introduced in [64] and that was called a "model with driving variables" in [67, Section 4.7]. Indeed, for a "driving variable representation" the matrix G has the specific form $G = [I \quad 0]$. The second representation (1.4) is in some sense dual to the first one; we will call this representation a "dual pencil (DP) representation". This type of representation has been

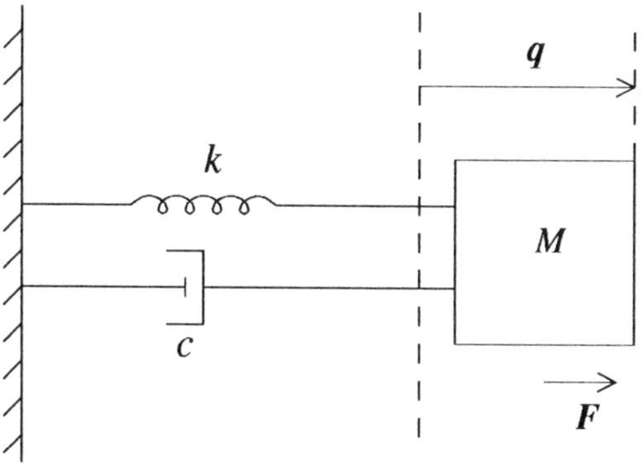

Figure 1.4: Mass-spring-damper system

considered before in [1, 68]. The DP representation is a more general form of the "output nulling state space representation" in [63, Section 3.5]. Indeed, for a "output nulling state space representation" the matrix K has the specific form $K = [I \quad 0]^T$.

Example 1.3 The mass-spring-damper system of Figure 1.4 is given by

$$\begin{bmatrix} M\frac{d^2}{dt^2} + c\frac{d}{dt} + k & -1 \end{bmatrix} \begin{bmatrix} q \\ F \end{bmatrix} = 0. \tag{1.5}$$

For the sake of simplicity, we will assume that $M = 1$. A P representation for the system is then given by:

$$\begin{aligned} \begin{bmatrix} 1 & 0 & 0 \\ 0 & 1 & 0 \end{bmatrix} \begin{bmatrix} \dot{\xi}_1 \\ \dot{\xi}_2 \\ \dot{\xi}_3 \end{bmatrix} &= \begin{bmatrix} -c & 1 & 0 \\ -k & 0 & 1 \end{bmatrix} \begin{bmatrix} \xi_1 \\ \xi_2 \\ \xi_3 \end{bmatrix} \\ \begin{bmatrix} q \\ F \end{bmatrix} &= \begin{bmatrix} 1 & 0 & 0 \\ 0 & 0 & 1 \end{bmatrix} \begin{bmatrix} \xi_1 \\ \xi_2 \\ \xi_3 \end{bmatrix}. \end{aligned} \tag{1.6}$$

A DP representation is:

$$\begin{bmatrix} 1 & 0 \\ 0 & 1 \\ 0 & 0 \end{bmatrix} \begin{bmatrix} \dot{\xi}_1 \\ \dot{\xi}_2 \end{bmatrix} = \begin{bmatrix} -c & 1 \\ -k & 0 \\ 1 & 0 \end{bmatrix} \begin{bmatrix} \xi_1 \\ \xi_2 \end{bmatrix} + \begin{bmatrix} 0 & 0 \\ 0 & 1 \\ -1 & 0 \end{bmatrix} \begin{bmatrix} q \\ F \end{bmatrix}. \tag{1.7}$$

$\diamond$

There are situations in which there is a natural decomposition of $\boldsymbol{w}$ in inputs and outputs. One might think of a controller as a system with a clear input-output decomposition: the inputs are the plant variables to be controlled (or measurements thereof), and the outputs are the control forces to be applied to the plant. In addition to the above representations, we will therefore consider a representation for which an a priori decomposition of $\boldsymbol{w}$ in inputs $\boldsymbol{u}$ and outputs $\boldsymbol{y}$ *is* required. The representation is given as

$$\sigma E\boldsymbol{\xi} = A\boldsymbol{\xi} + B\boldsymbol{u} \tag{1.8}$$

$$\boldsymbol{y} = C\boldsymbol{\xi} + D\boldsymbol{u} \tag{1.9}$$

and has been studied extensively in the literature (see e.g. [39] and references therein); it is usually called a "descriptor representation", a terminology that stems from Luenberger [41]. In the sequel we will call (1.8–1.9) a "D representation".

For us there are two kinds of motivations to consider D representations. First, there are situations in which the descriptor form arises naturally from certain laws that underlie the system. The next example is drawn from economics; see [8, 40].

Example 1.4 Consider an economic process that involves n interrelated production sectors. The relationships between the levels of production of the sectors can be described by a so-called Leontieff model:

$$\boldsymbol{x}(k) = A\boldsymbol{x}(k) + B(\boldsymbol{x}(k+1) - \boldsymbol{x}(k)) + \boldsymbol{d}(k). \tag{1.10}$$

Here the components of the n-dimensional vector $\boldsymbol{x}(k)$ are the levels of production of the sectors at time k. The vector $A\boldsymbol{x}$ should be interpreted as the capital that is required as direct input for production of $\boldsymbol{x}$; a coefficient a_{ij} of the "flow coefficient matrix" A indicates the amount of product i that is needed to produce one unit of product j. The vector $B\boldsymbol{x}$ stands for the capital that is required to be in stock to be able to produce $\boldsymbol{x}$ in the next time period. A coefficient b_{ij} of the "stock coefficient matrix" B indicates the amount of product i that has to be in stock to be able to produce one unit of product j in the next time period. The vector $\boldsymbol{d}(k)$ represents the levels of production that are demanded. Econometric models of this type, both in discrete time and in continuous time (see also [49]), were considered by Leontieff in [38].

It often happens that production in one sector doesn't need capital in stock from all the other sectors. Moreover, in practice there are usually few sectors that offer capital in stock to other sectors (for instance, agriculture does not produce capital in stock). As a result, most of the elements in the stock coefficient matrix B are zero and B is often singular. The representation (1.10) can be rewritten in descriptor form as

$$B\boldsymbol{x}(k+1) = (I - A + B)\boldsymbol{x}(k) - \boldsymbol{d}(k)$$

and is thus an example of how a descriptor form can arise in practical situations as a natural description of a dynamical system. ◇

The electrical circuit of Example 1.1 gives rise to another situation in which a descriptor representation arises naturally:

Example 1.5 Let us introduce an additional variable $\boldsymbol{V}_2$ in the electrical circuit of Example 1.1, so that the circuit becomes as in the following figure:

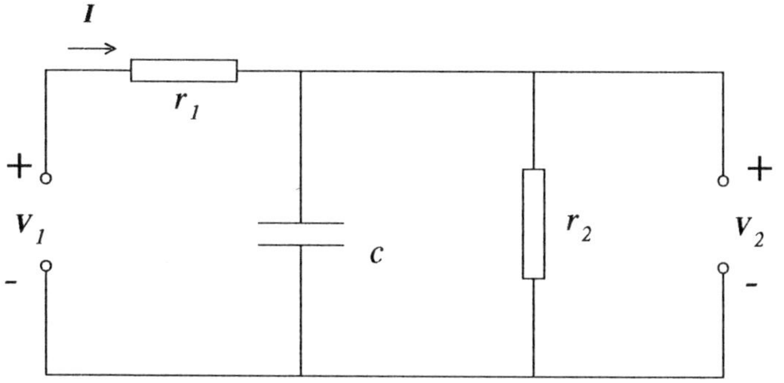

Let us consider the voltage $\boldsymbol{V}_1$ as $\boldsymbol{u}$ and the voltage $\boldsymbol{V}_2$ as $\boldsymbol{y}$. Taking $\boldsymbol{V}_2$ and the current intensity $\boldsymbol{I}$ as the descriptor variables ξ_1 and ξ_2 respectively, we can represent the system in descriptor form by

$$\begin{aligned} \begin{bmatrix} cr_2 & 0 \\ 0 & 0 \end{bmatrix} \begin{bmatrix} \dot{\xi}_1 \\ \dot{\xi}_2 \end{bmatrix} &= \begin{bmatrix} -1 & r_2 \\ 1 & r_1 \end{bmatrix} \begin{bmatrix} \xi_1 \\ \xi_2 \end{bmatrix} + \begin{bmatrix} 0 \\ -1 \end{bmatrix} \boldsymbol{u} \\ \boldsymbol{y} &= [1 \quad 0] \begin{bmatrix} \xi_1 \\ \xi_2 \end{bmatrix}. \end{aligned} \tag{1.11}$$

For $c > 0$, $r_1 > 0$ and $r_2 > 0$, we may rewrite this as the standard state space representation

$$\begin{aligned} \dot{\boldsymbol{x}} &= -\frac{r_1 + r_2}{cr_1 r_2}\boldsymbol{x} + \frac{1}{cr_1}\boldsymbol{u} \\ \boldsymbol{y} &= \boldsymbol{x}. \end{aligned} \tag{1.12}$$

Note that the parameters in (1.11) appear explicitly, whereas additional computations are required for (1.12). Note further that this example presents a descriptor representation that can be written in standard state space form despite the fact that the matrix E is singular. In Chapter 6 we will give conditions on E, A, B, C, and D under which the D representation (E, A, B, C, D) can be rewritten in standard state space form. We will also give the procedure. ◇

Other examples where the descriptor form arises naturally are related to systems that are an interconnection of subsystems. The interconnection gives rise to algebraic constraints on the relevant variables, and these can be directly expressed by means of a D representation. Indeed, in case that the matrix E is singular, the dynamical equation (1.8) of a D representation consists of a combination of differential (or difference) equations and

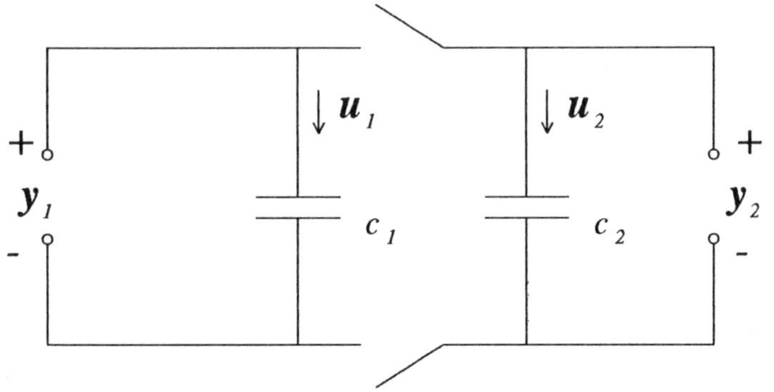

Figure 1.5: Electrical circuit II

algebraic equations. In this context, the matrices E and A in the D representation are typically nonsquare. Applications can be found, for instance, in the modeling of electrical circuits, as is shown by the following example. In fact, the ideas of the next example can be extended to any RCL electrical network, since such a network is made up of linear first-order differential equations and linear static equations.

Example 1.6 Consider two electrical circuits that can be interconnected by switches as in Figure 1.5. We will assume that the capacitance c_2 equals 1 and that $c_1 > 0$. As indicated in the figure, we consider the currents as inputs ($\boldsymbol{u}_1$ and $\boldsymbol{u}_2$) and the voltages as outputs ($\boldsymbol{y}_1$ and $\boldsymbol{y}_2$). When the switches are open, a representation of the system is given by

$$\begin{bmatrix} \dot{\boldsymbol{x}}_1 \\ \dot{\boldsymbol{x}}_2 \end{bmatrix} = \begin{bmatrix} 1/c_1 & 0 \\ 0 & 1 \end{bmatrix} \begin{bmatrix} \boldsymbol{u}_1 \\ \boldsymbol{u}_2 \end{bmatrix}$$
$$\begin{bmatrix} \boldsymbol{y}_1 \\ \boldsymbol{y}_2 \end{bmatrix} = \begin{bmatrix} 1 & 0 \\ 0 & 1 \end{bmatrix} \begin{bmatrix} \boldsymbol{x}_1 \\ \boldsymbol{x}_2 \end{bmatrix}.$$

Closing the switches imposes the additional constraint

$$\boldsymbol{y}_1 = \boldsymbol{y}_2. \tag{1.13}$$

The system is then described by the D representation

$$\begin{bmatrix} 1 & 0 \\ 0 & 1 \\ 0 & 0 \end{bmatrix} \begin{bmatrix} \dot{\boldsymbol{x}}_1 \\ \dot{\boldsymbol{x}}_2 \end{bmatrix} = \begin{bmatrix} 0 & 0 \\ 0 & 0 \\ 1 & -1 \end{bmatrix} \begin{bmatrix} \boldsymbol{x}_1 \\ \boldsymbol{x}_2 \end{bmatrix} + \begin{bmatrix} 1/c_1 & 0 \\ 0 & 1 \\ 0 & 0 \end{bmatrix} \begin{bmatrix} \boldsymbol{u}_1 \\ \boldsymbol{u}_2 \end{bmatrix}$$
$$\begin{bmatrix} \boldsymbol{y}_1 \\ \boldsymbol{y}_2 \end{bmatrix} = \begin{bmatrix} 1 & 0 \\ 0 & 1 \end{bmatrix} \begin{bmatrix} \boldsymbol{x}_1 \\ \boldsymbol{x}_2 \end{bmatrix}. \tag{1.14}$$

After closing the switches, the system can alternatively be represented with fewer $\boldsymbol{x}$-variables by the D representation

$$\begin{aligned}\begin{bmatrix}1\\0\end{bmatrix}\dot{\boldsymbol{x}} &= \begin{bmatrix}0\\0\end{bmatrix}\boldsymbol{x}+\begin{bmatrix}0 & 1\\1 & -c_1\end{bmatrix}\begin{bmatrix}\boldsymbol{u}_1\\\boldsymbol{u}_2\end{bmatrix}\\ \begin{bmatrix}\boldsymbol{y}_1\\\boldsymbol{y}_2\end{bmatrix} &= \begin{bmatrix}1\\1\end{bmatrix}\boldsymbol{x}.\end{aligned} \tag{1.15}$$

We will see in Chapter 4 that the above representation is a minimal D representation of this system. Note that the variables $\boldsymbol{u}_1$ and $\boldsymbol{u}_2$ can *not* be considered as inputs for the interconnected system, since they are related through $\boldsymbol{u}_1 = c_1\boldsymbol{u}_2$. ◇

If the matrix E is invertible, a D representation can, of course, be rewritten in the well-known standard state space form:

$$\begin{aligned}\sigma\boldsymbol{x} &= A\boldsymbol{x}+B\boldsymbol{u}\\ \boldsymbol{y} &= C\boldsymbol{x}+D\boldsymbol{u}.\end{aligned} \tag{1.16}$$

However, if the D representation is derived from first principles, as in the above examples, it displays the system parameters exactly as they appear in the basic dynamical equations. For numerical reasons, then, there are advantages to consider the descriptor form rather than the standard state space form. This holds especially true if E is a matrix that is ill-conditioned. In the same way large scale systems may give rise to D representations with large and sparse matrices. For the analysis of such systems one might want to apply graph-theoretical techniques, and it is then preferable to deal directly with the descriptor form as in [46].

Our second motivation to consider D representations is that these allow for the representation of systems in which the inputs and outputs are related by a nonproper transfer function. This is unlike the situation for standard state space representations (1.16), which only give rise to proper rational transfer functions. In practice there are several situations in which it is common to consider the inputs and outputs as being related by a nonproper transfer function. We mention the following example.

Example 1.7 A controller that is often used in industry is the so-called PD (Proportional-Derivative) controller. A model for this controller is given by differential equations of the type

$$\boldsymbol{F}(t) = K_p\boldsymbol{q}(t) + K_d\dot{\boldsymbol{q}}(t). \tag{1.17}$$

Here $\boldsymbol{q}$ denotes the plant variables to be controlled, $\boldsymbol{F}$ denotes the control variables to be applied to the plant and K_p and K_d are constant gain matrices. As usual, we consider $\boldsymbol{q}$ as the input vector and $\boldsymbol{F}$ as the output vector. Input and output are then related by the transfer function K_p+sK_d, which is not proper. As a result, the controller cannot be represented in standard state space form. However, a D representation is possible: the system is represented by

$$\begin{aligned}
\begin{bmatrix} 0 & 1 \\ 0 & 0 \end{bmatrix}\begin{bmatrix} \dot{\xi}_1 \\ \dot{\xi}_2 \end{bmatrix} &= \begin{bmatrix} 1 & 0 \\ 0 & 1 \end{bmatrix}\begin{bmatrix} \xi_1 \\ \xi_2 \end{bmatrix} + \begin{bmatrix} 0 \\ -1 \end{bmatrix} \boldsymbol{q} \\
\boldsymbol{F} &= [K_d \quad K_p]\begin{bmatrix} \xi_1 \\ \xi_2 \end{bmatrix}.
\end{aligned} \tag{1.18}$$

◇

Other examples of systems with nonproper transfer functions are systems in which the variables are functions of space rather than of time, in particular n-D systems; see e.g. [37, 50]. These systems, for instance, are relevant for situations in image processing and seismology. We believe that a good understanding of the nonproper 1-D case can be useful for research in n-D systems.

The outline of the book is as follows.

In Chapter 2 we present some basic concepts and results, which are essential to the development in the remaining chapters. The first part of Chapter 2 is concerned with algebraic theory for complex-rational matrices. Results for real-rational matrices are then immediate and will be used in subsequent chapters. Following Verghese and co-workers [56, 61], we introduce the notion of the "content" (pole/zero excess) of a rational matrix. So-called "Wiener-Hopf indices" of rational matrices are also introduced in this part. As pointed out in [56, 61], the notion of "content" is useful in connection with the results of Forney [15] concerning polynomial bases of rational vector spaces. We find that the combination of this notion and the "Wiener-Hopf indices" is especially powerful in this context. The second part of Chapter 2 deals with the geometric theory of matrix pencils. Most of the results in this part are known, but we present a new result in connecting the rank of a constant matrix to properties of a related matrix pencil and related rational vector spaces.

In Chapter 3 dynamical systems are formally defined. We consider both the discrete-time case $\mathbf{T} = \mathbb{Z}_+$ and the continuous-time case $\mathbf{T} = \mathbb{R}_+$; the subsequent development allows a unified treatment of these cases. In the first part of the chapter, we introduce a class of representations involving exclusively external variables. We then define several structural properties in terms of a representation of this type. This includes the definition of observability indices and controllability indices, notions that are classically defined at the much less general level of standard state space representations. Subsequently, we define a class of representations in which not all variables have to be external variables. Particular members of this class are the abovementioned P representation, DP representation, and D representation. Most results in this part of the chapter originate from the work of Willems [66, 67, 68]. We do, however, present a new result in deriving a space of rational vectors that is a complete invariant under the equivalence relation that is induced by the behavioral approach. Expressions for this space in terms of the types of representations considered before are given. Next, we introduce a weaker type of equivalence that proves useful in formulating duality results. This type of equivalence is actually a more general version of the well-known notion of transfer equivalence, according to which systems are equivalent if their transfer functions are the same. It coincides with equivalence notions in the work of Aplevich [1]–[3] and Grimm [21]. In the second part of Chapter 3, we investigate the relation between the pencil form and the descriptor form.

Central questions of Chapter 4 are concerned with minimality and transformation groups of the first-order representations mentioned earlier. The abovementioned result of Chapter 2 on the rank of a constant matrix is put to work in the first part of the chapter. By this result invariant lower bounds for *all* items to be minimized can be derived in a direct way; except for standard state space representations, such a result has not been established before.

In Chapter 5 we study methods for obtaining minimal first-order realizations. The work of Willems [66] and Fuhrmann [16, 18] is unified in this chapter. The basic realization is obtained in pencil form; realizations in dual pencil form and descriptor form are deduced from the basic realization.

Chapter 6 is concerned with the expression of certain structural invariants in terms of the above types of first-order representations. We derive explicit formulas in terms of the original parameters of representations with an arbitrary amount of redundancy. In the first part of the chapter, we give expressions for the observability indices and controllability indices mentioned above. The second part deals with questions concerning the input/output status of external variables. We give characterizations in terms of the various types of first-order representations. In the same framework we derive expressions for the rank and pole/zero structure at infinity of the transfer function of an input-output system. We include a procedure to derive a standard state space description from a given descriptor representation. Throughout the chapter an important role is played by the Kronecker invariants of a matrix pencil.

2

Rational matrices and vector spaces

2.1 Algebraic preliminaries

In this section we summarize some basic algebraic definitions and results that will be needed throughout this book. For more details the reader is referred to textbooks on matrices over rings such as [47]. We assume that the reader is familiar with the notion of a field as a collection of objects that can be added, subtracted, multiplied and divided with the usual associative, distributive and commutative rules. The fields of importance in this book are the field of real numbers $\mathbb{R}$, the field of complex numbers $\mathbb{C}$ and the field of rational functions in s with coefficients in $\mathbb{C}$ ($\mathbb{R}$), written conventionally as $\mathbb{C}(s)$ ($\mathbb{R}(s)$). We also assume that the reader is familiar with the notion of a commutative ring as a collection of objects with all the properties of a field except division. The set of polynomials in s with coefficients in $\mathbb{C}$ (or $\mathbb{R}$) is a commutative ring, written conventionally as $\mathbb{C}[s]$ ($\mathbb{R}[s]$).

A commutative ring $\mathcal{R}$ which has an identity element and no zero divisors is called an *integral domain*. The elements in $\mathcal{R}$ that have inverses are called *units*. A *principal ideal domain* is a special type of integral domain. It has the convenient property that each element can be written as a finite product of irreducible elements, i.e., elements that have no nontrivial factors; the factorization is unique up to order and units. The notion of *greatest common divisor* (*least common multiple*) of a finite set of elements as the greatest (smallest) ring element that is a divisor (multiple) of all elements of the set is well defined in a principal ideal domain. Apart from the above factorization property a principal ideal domain $\mathcal{R}$ also has the so-called *Bézout property*: there exist $a \in \mathcal{R}$ and $b \in \mathcal{R}$ such that the greatest common divisor d of two nonzero elements f and g can be written as

$$d = fa + gb.$$

An integral domain $\mathcal{R}$ is called a *euclidean domain* if there exists a function $v : \mathcal{R} \setminus \{0\} \to \mathbb{Z}_+$ (*euclidean function*) with the following properties:

(a) $v(fg) \geq v(f)$ for all nonzero f and g, and

(b) for all nonzero f and g there exist p and r such that $f = gp + r$ with either $v(r) < v(g)$ or $r = 0$ (*division with remainder*).

Not every principal ideal domain is a euclidean domain; however, a euclidean domain is always a principal ideal domain. The ring $\mathbb{C}[s]$ is a euclidean domain: the degree function serves as a euclidean function. In the next section we will see that it is not the only euclidean domain in $\mathbb{C}(s)$.

For an integral domain $\mathcal{R}$ the set of $n \times n$-matrices over $\mathcal{R}$, denoted by $\mathcal{R}^{n \times n}$, is a ring which has an identity element. The units in $\mathcal{R}^{n \times n}$ are the matrices for which the determinant is a unit in $\mathcal{R}$. Such matrices will be called *$\mathcal{R}$-unimodular*. A matrix in $\mathcal{R}^{n \times m}$ will be called *left (right) $\mathcal{R}$-unimodular* if it has a left (right) inverse in $\mathcal{R}^{m \times n}$.

For matrices over a principal ideal domain $\mathcal{R}$ we have the following theorem that is proved in, for instance, [47, Theorem II.9]:

Theorem 2.1 (*Invariant Factor Theorem*) Let $\mathcal{R}$ be a principal ideal domain and let T be an $n \times m$-matrix over $\mathcal{R}$. Then T can be factorized as

$$T = UDV \text{ with } D = \begin{bmatrix} \Delta & 0 \\ 0 & 0 \end{bmatrix} \tag{2.1}$$

where U and V are $\mathcal{R}$-unimodular matrices of sizes $n \times n$ and $m \times m$ respectively; $\Delta = \text{diag}\,(d_1, d_2, \ldots, d_r)$ with $d_i \in \mathcal{R}$, and d_i divides d_{i+1}. Moreover, the elements d_i are unique up to units ($i = 1, \ldots, r$). ◇

The canonical form D in (2.1) is commonly referred to as the *Smith form over $\mathcal{R}$* of the matrix T.

The *quotient field* of an integral domain $\mathcal{R}$ is defined by means of the following equivalence relation on the product $\mathcal{R} \times (\mathcal{R} \setminus \{0\})$:

$$(f_1, g_1) \text{ is equivalent to } (f_2, g_2) \quad :\Longleftrightarrow \quad f_1 g_2 = f_2 g_1.$$

The equivalence class that corresponds to the pair $(f, g) \in \mathcal{R} \times (\mathcal{R} \setminus \{0\})$ is usually written as the quotient f/g (note, however, that g need not be a unit in $\mathcal{R}$). In a trivial way $\mathcal{R}$ can be considered as a subring of its quotient field. The extension of the previous theorem to matrices over quotient fields is immediate.

Theorem 2.2 (*Extended Invariant Factor Theorem*) Let $\mathcal{F}$ be the quotient field of a principal ideal domain $\mathcal{R}$, and let T be an $n \times m$-matrix over $\mathcal{F}$. Then T can be factorized as

$$T = UDV \text{ with } D = \begin{bmatrix} \Delta & 0 \\ 0 & 0 \end{bmatrix}$$

where U and V are $\mathcal{R}$-unimodular matrices of sizes $n \times n$ and $m \times m$, respectively; $\Delta = \text{diag}\ (d_1, d_2, \ldots, d_r)$ with $d_i \in \mathcal{F}$ and $d_{i+1}/d_i \in \mathcal{R}$. Moreover, the elements d_i are unique up to units in $\mathcal{R}$ $(i = 1, \ldots, r)$.

Proof (cf. [43, 14]) Let f be the least common multiple of the denominators of the elements of T. Then fT is a matrix over $\mathcal{R}$, so that we may apply Theorem 2.1, leading to $fT = U\tilde{D}V$. Defining $D = \tilde{D}/f$ we get the desired factorization. ◇

The above two theorems are well-known in linear system theory. In former times their use was confined to rings of polynomials; see, for instance, the work of McMillan [43], Kalman [27] and Rosenbrock [51]. The utility of considering other rings of rational functions as well has first been recognized by Hautus [22], Morse [45] and Verghese et al. [56, 58]. In the next section we will introduce these "other rings" in a rather general setting in which no point in the extended complex plane is assigned any special role a priori.

2.2 Euclidean domains of rational functions

In the sequel we denote the extended complex plane $\mathbb{C} \cup \{\infty\}$ by $\mathbb{C}^e$. Throughout the section we assume that $\alpha \in \mathbb{C}^e$ is arbitrary, unless stated otherwise. Following Verghese and co-workers [56, 61] we have the following definition:

Definition 2.3 Let $f(s) \in \mathbb{C}(s)$ be nonzero. The *content at* α of $f(s)$, denoted by $c_\alpha(f)$, is defined as the pole multiplicity at α of $f(s)$ (where a zero of multiplicity k is counted as a pole of multiplicity $-k$). ◇

For nonzero $f(s)$ and $g(s) \in \mathbb{C}(s)$ we have the following properties (which show that the function $c_\alpha(.)$ is the negative of a so-called "discrete valuation" on $\mathbb{C}(s)$; see [5]):

Property 2.1 $\sum_{\alpha \in \mathbb{C}^e} c_\alpha(f) = 0$

Property 2.2 $c_\alpha(fg) = c_\alpha(f) + c_\alpha(g)$

Property 2.3 $c_\alpha(f+g) \leq \max\ (c_\alpha(f), c_\alpha(g))$

A generalization of the Euclidean domain $\mathbb{C}[s]$ is provided by the following definition.

Definition 2.4 $\mathbb{C}_\alpha^+(s) = \{f(s) \in \mathbb{C}(s) \mid c_\beta(f) \leq 0 \text{ for all } \beta \in \mathbb{C}^e \setminus \{\alpha\}\}$

The set $\mathbb{C}_\infty^+(s)$ coincides with $\mathbb{C}[s]$; for a polynomial $f(s) \in \mathbb{C}_\infty^+(s)$ the value of $c_\infty(f)$ equals the degree of $f(s)$. In general, $\mathbb{C}_\alpha^+(s)$ is a euclidean domain with the content at α serving as a euclidean function. The elements of

$\mathbb{C}_\alpha^+(s)$ are finite linear combinations of nonpositive powers of the function $(s-\alpha)$; the units in $\mathbb{C}_\alpha^+(s)$ are the constants.

Next we define a different type of euclidean domain in $\mathbb{C}(s)$.

Definition 2.5 $\mathbb{C}_\alpha^-(s) = \{f(s) \in \mathbb{C}(s) \mid c_\alpha(f) \leq 0\}$

The set $\mathbb{C}_\infty^-(s)$ is the ring of proper rational functions; for $f(s) \in \mathbb{C}_\infty^-(s)$ the value of $c_\infty(f)$ is the difference between the degrees of the denominator and the numerator polynomials of $f(s)$. In general, $\mathbb{C}_\alpha^-(s)$ is a euclidean domain with the negative of the content at α serving as a euclidean function (see, for instance, the proof of Theorem 2.1.4 in [62]). The units in $\mathbb{C}_\alpha^-(s)$ are the elements $f(s)$ for which $c_\alpha(f) = 0$ or, equivalently, for which $f(\alpha) \neq 0$. Note that we have

$$\mathbb{C}_\alpha^+(s) = \{f(s) \in \mathbb{C}(s) \mid f(s) \in \mathbb{C}_\beta^-(s) \text{ for all } \beta \in \mathbb{C}^e \setminus \{\alpha\}\} \tag{2.2}$$

The field $\mathbb{C}(s)$ is, by definition, the quotient field of $\mathbb{C}_\infty^+(s)$; from this it follows immediately that $\mathbb{C}(s)$ is the quotient field of $\mathbb{C}_\alpha^+(s)$ for all $\alpha \in \mathbb{C}^e$. It can be easily shown that $\mathbb{C}(s)$ is also the quotient field of $\mathbb{C}_\alpha^-(s)$ for all $\alpha \in \mathbb{C}^e$, which is a property that we will make use of in the next section.

2.3 Pole/zero structure of a rational matrix

Throughout the section we assume $\alpha \in \mathbb{C}^e$ to be arbitrary. We first take $\mathcal{R} = \mathbb{C}_\alpha^-(s)$ in Theorem 2.2 to obtain a canonical form for complex-rational matrices (analogous results for real-rational matrices are then obvious). Note that an element $d_{i+1}(s)/d_i(s) \in \mathbb{C}(s)$ belongs to $\mathbb{C}_\alpha^-(s)$ if and only if $c_\alpha(d_i) \geq c_\alpha(d_{i+1})$.

Theorem 2.6 Every matrix $T(s) \in \mathbb{C}^{n \times m}(s)$ of rank r can be factorized as

$$T(s) = U(s)D(s)V(s) \text{ with } D(s) = \begin{bmatrix} \Delta(s) & 0 \\ 0 & 0 \end{bmatrix} \tag{2.3}$$

where $U(s)$ and $V(s)$ are $\mathbb{C}_\alpha^-(s)$-unimodular matrices of sizes $n \times n$ and $m \times m$ respectively; $\Delta(s) = \text{diag}\,(d_1(s), d_2(s), \ldots, d_r(s))$ with $c_\alpha(d_i) \geq c_\alpha(d_{i+1})$. Moreover, the rational functions $d_i(s)$ are unique up to units in $\mathbb{C}_\alpha^-(s)$ $(i = 1, \ldots, r)$. ◇

In accordance with [61] we will call the canonical form $D(s)$ in (2.3) the *Smith-McMillan form at* α of $T(s)$. The integers $c_\alpha(d_1), \ldots, c_\alpha(d_r)$ will be called the *orders of the poles at* α *of* $T(s)$ (poles of order $-k$ are also called *zeros* of order k). The *content at* α of $T(s)$, denoted by $c_\alpha(T)$, is defined as

$$c_\alpha(T) := \sum_{i=1}^{r} c_\alpha(d_i).$$

The *pole multiplicity at* α of $T(s)$, denoted by $\delta^p_\alpha(T)$, is defined as

$$\delta^p_\alpha(T) := \sum_{\{i \mid c_\alpha(d_i)>0\}} c_\alpha(d_i).$$

The *zero multiplicity at* α of $T(s)$, denoted by $\delta^z_\alpha(T)$, is defined as

$$\delta^z_\alpha(T) := -\sum_{\{i \mid c_\alpha(d_i)<0\}} c_\alpha(d_i).$$

Note that, by definition, $c_\alpha(T) = \delta^p_\alpha(T) - \delta^z_\alpha(T)$. We conclude our list of definitions with the following definition, which stems from [56, 61]:

Definition 2.7 Let $T(s)$ be a matrix over $\mathbb{C}(s)$. The *content* of $T(s)$, denoted by $c(T)$, is defined as

$$c(T) = \sum_{\alpha\in\mathbb{C}^e} c_\alpha(T).$$

The *total pole multiplicity* of $T(s)$, denoted by $\delta^p(T)$, is defined as

$$\delta^p(T) = \sum_{\alpha\in\mathbb{C}^e} \delta^p_\alpha(T).$$

The *total zero multiplicity* of $T(s)$, denoted by $\delta^z(T)$, is defined as

$$\delta^z(T) = \sum_{\alpha\in\mathbb{C}^e} \delta^z_\alpha(T).$$

◇

Note that $\delta^p(T)$ coincides with the well-known *McMillan degree* of $T(s)$, as it was introduced by McMillan in [43].

The following lemma will be very useful in the sequel.

Lemma 2.8 Let $T(s) \in \mathbb{C}^{n\times m}(s)$ be a matrix of rank r. Let $n_1,\ldots,n_r$ be the orders of the poles at α of $T(s)$. Define, for $k \in \mathbb{Z}$,

$$p_k := \sharp\{i \in \{1,\ldots,r\} \mid n_i \geq k\}$$

and

$$s_k := \sharp\{i \in \{1,\ldots,r\} \mid n_i \leq k\}.$$

Then

$$p_k = \dim\{g(\alpha) \mid g(s) \in \mathbb{C}^-_\alpha(s) \text{ and } g(s) = (s-\alpha)^k T(s) f(s) \text{ for some } f(s) \in \mathbb{C}^-_\alpha(s)\} \tag{2.4}$$

and

$$s_k = \dim\{f(\alpha) \mid f(s) \in \mathbb{C}^-_\alpha(s) \text{ and } (s-\alpha)^k T(s) f(s) \in \mathbb{C}^-_\alpha(s)\} - \dim\ \ker T(s). \tag{2.5}$$

Proof (cf. [48]) The right-hand sides of both (2.4) and (2.5) are invariant under left and right multiplication by $\mathbb{C}_\alpha^-(s)$-unimodular matrices. This can be easily verified by noting that a matrix $U(s)$ is $\mathbb{C}_\alpha^-(s)$-unimodular if and only if $U(\alpha)$ is invertible. We may therefore assume that $T(s)$ is in the form $D(s)$ in (2.3) (Smith-McMillan form at α) and then (2.4) and (2.5) follow immediately. ◇

In the next lemma (see [56, 61]) the pole structure of a matrix T is expressed in terms of its nonzero minors; here a j-minor of T is defined as the determinant of a $j \times j$-submatrix of T. The expression turns out to be a useful tool in the sequel.

Lemma 2.9 ([56]) Let $T(s) \in \mathbb{C}^{n\times m}(s)$ be a matrix of rank $r \geq 1$. Let $n_1 \geq n_2 \geq \cdots \geq n_r$ be the orders of the poles at α of $T(s)$. Then we have, for $1 \leq j \leq r$,

$$\sum_{i=1}^{j} n_i = \max\{c_\alpha(f) \mid f(s) \text{ is a nonzero } j\text{-minor of } T(s)\}. \tag{2.6}$$

Proof We first prove that the right-hand side of (2.6) is invariant under left multiplication of $T(s)$ by $\mathbb{C}_\alpha^-(s)$-unimodular matrices. For this, let $T_2(s) = T_1(s)T(s)$ where $T_1(s)$ is a $g \times n$-matrix over $\mathbb{C}_\alpha^-(s)$ that has full column rank. Let $j \in \{1, \ldots, r\}$ and define

$$M_1 := \max\{c_\alpha(p) \mid p(s) \text{ is a nonzero } j\text{-minor of } T_1(s)\}$$

and

$$M_2 := \max\{c_\alpha(q) \mid q(s) \text{ is a nonzero } j\text{-minor of } T_2(s)\}.$$

Denoting the right-hand side of (2.6) by M, we conclude from Property 2.2 and the Binet-Cauchy formula (see, for instance, [47, Chapter II.12]) that

$$M_2 \leq M_1 + M \leq M,$$

where the last inequality holds because of the fact that $T_1(s)$ is a matrix over $\mathbb{C}_\alpha^-(s)$. In case that $T_1(s)$ is $\mathbb{C}_\alpha^-(s)$-unimodular we may also write

$$T(s) = T_1^{-1}(s)T_2(s),$$

and we conclude from the same argument that $M \leq M_2$. This yields $M = M_2$ so that the right-hand side of (2.6) is indeed invariant under left multiplication of $T(s)$ by $\mathbb{C}_\alpha^-(s)$-unimodular matrices. By a similar argument it is also invariant under right multiplication of $T(s)$ by $\mathbb{C}_\alpha^-(s)$-unimodular matrices. We may therefore assume that $T(s)$ is in Smith-McMillan form at α, and then (2.6) follows immediately. ◇

The content of a scalar rational function is always zero (Property 2.1). In the matrix case we have a different situation:

Lemma 2.10 ([56, 61]) Let $T(s) \in \mathbb{C}^{n\times m}(s)$. Then

$$c(T) \geq 0$$

with equality if and only if $T(s)$ is nonsingular.

Proof Denote the rank of $T(s)$ by r. Let $f(s)$ be an r-minor of $T(s)$. By Lemma 2.9 we have that

$$c(T) = \sum_{\alpha\in\mathbb{C}^e} c_\alpha(T) \geq \sum_{\alpha\in\mathbb{C}^e} c_\alpha(f) = 0. \tag{2.7}$$

If $T(s)$ is nonsingular then $f(s) = \det T(s)$ is the only r-minor of $T(s)$. As a result $c_\alpha(T) = c_\alpha(f)$ for all $\alpha \in \mathbb{C}^e$ so that equality holds in (2.7). ◇

The following lemma is the analog of Property 2.2.

Lemma 2.11 Let $T(s) \in \mathbb{C}^{n\times m}(s)$ be a matrix of rank r. Let $T(s)$ be factorized as

$$T(s) = T_1(s)T_2(s)$$

where $T_1(s) \in \mathbb{C}^{n\times r}(s)$ has full column rank and $T_2(s) \in \mathbb{C}^{r\times m}(s)$ has full row rank. Then for all $\alpha \in \mathbb{C}^e$

$$c_\alpha(T) = c_\alpha(T_1) + c_\alpha(T_2).$$

In particular we have that

$$c(T) = c(T_1) + c(T_2).$$

Proof (cf. [56, 61]) It follows from the proof of Lemma 2.9 that $c_\alpha(T) \leq c_\alpha(T_1) + c_\alpha(T_2)$. We will now prove that also $c_\alpha(T) \geq c_\alpha(T_1) + c_\alpha(T_2)$, from which the desired result follows. Let $f_1(s)$ be a nonzero r-minor of $T_1(s)$ for which the content at α is maximal, and let $f_2(s)$ be a nonzero r-minor of $T_2(s)$ for which the content at α is maximal. Because of the rank assumptions, the Binet-Cauchy formula implies that $f(s) = f_1(s)f_2(s)$ is a nonzero r-minor of $T(s)$ so that $c_\alpha(T_1) + c_\alpha(T_2) = c_\alpha(f_1) + c_\alpha(f_2) = c_\alpha(f) \leq c_\alpha(T)$. ◇

We are now able to characterize $\mathbb{C}_\alpha^-(s)$-unimodularity and $\mathbb{C}_\alpha^+(s)$-unimodularity in a concise way. The following lemmas present well-known results and can be easily deduced from Lemma 2.9 and (2.2).

Lemma 2.12 Let $T(s)$ be a matrix over $\mathbb{C}_\alpha^-(s)$ of rank r. Then $c_\alpha(T) \leq 0$ with equality if and only if $T(\alpha)$ has rank r. In particular, $T(s)$ is left (right) $\mathbb{C}_\alpha^-(s)$-unimodular if and only if $T(\alpha)$ has full column (row) rank.

Lemma 2.13 Let $T(s)$ be a matrix over $\mathbb{C}_\alpha^+(s)$ of rank r. Let $\beta \in \mathbb{C}^e \backslash \{\alpha\}$. Then $c_\beta(T) \leq 0$ with equality if and only if $T(\beta)$ has rank r. In particular, $T(s)$ is left (right) $\mathbb{C}_\alpha^+(s)$-unimodular if and only if $T(\beta)$ has full column (row) rank for all $\beta \in \mathbb{C}^e \setminus \{\alpha\}$. ◇

In system theory one often deals with matrices $T(s)$ that are factorized as $T(s) = D^{-1}(s)N(s)$ or as $T(s) = N(s)D^{-1}(s)$. In the next theorem properties of $T(s)$ are related to properties of the factors $D(s)$ and $N(s)$. We first present a lemma.

Lemma 2.14 Let $T(s) \in \mathbb{C}^{n\times m}(s)$.

$$\delta_\alpha^p(T) = c_\alpha([I_{n\times n} \quad T]) = c_\alpha(\begin{bmatrix} T \\ I_{m\times m} \end{bmatrix}).$$

In particular we have that

$$\delta^p(T) = c([I_{n\times n} \quad T]) = c(\begin{bmatrix} T \\ I_{m\times m} \end{bmatrix}). \tag{2.8}$$

Proof Denote the rank of $T(s)$ by r. It is not difficult to see that

$$\{f(s) \mid f(s) \text{ is a nonzero } n\text{-minor of } [I_{n\times n} \quad T]\} =$$

$$= \{1\} \cup \bigcup_{i=1}^{r} \{g(s) \mid g(s) \text{ is a nonzero } i\text{-minor of } T(s)\}.$$

Taking the maximum content at α on both sides yields $c_\alpha([I_{n\times n} \quad T]) = \delta_\alpha^p(T)$ (use Lemma 2.9). It is proven in a completely similar way that

$$c_\alpha(\begin{bmatrix} T \\ I_{m\times m} \end{bmatrix}) = \delta_\alpha^p(T),$$

and (2.8) follows immediately. ◇

A proof of the second equality in (2.8) is essentially present in [15] where the n-minors of $[I_{n\times n} \quad T]$ are brought into 1-1 correspondence with the m-minors of the matrix

$$\begin{bmatrix} T \\ I_{m\times m} \end{bmatrix}.$$

However, a direct connection with $\delta^p(T)$ is not established in the paper.

The results of the next two theorems are not new; they are essentially proven in [25, 59]. Nevertheless, we give proofs in order to demonstrate that the results can be obtained from the previous lemma in an elegant way.

Theorem 2.15 Let $T(s) \in \mathbb{C}^{n\times m}(s)$ be given as $T(s) = D^{-1}(s)N(s)$, where $D(s) \in \mathbb{C}^{n\times n}(s)$ and $N(s) \in \mathbb{C}^{n\times m}(s)$. Then

$$\delta^p(T) = c([D \quad N]). \tag{2.9}$$

Moreover, the following implications hold:

(i) $[D(s) \ \ N(s)]$ is right $\mathbb{C}^-_\alpha(s)$-unimodular $\Longrightarrow$ $\delta^p_\alpha(T) = -c_\alpha(\det\ D)$

(ii) $[D(s) \ \ N(s)]$ is right $\mathbb{C}^+_\alpha(s)$-unimodular $\Longrightarrow$

$$\sum_{\beta\in\mathbb{C}^e\setminus\{\alpha\}} \delta^p_\beta(T) = c_\alpha(\det\ D).$$

Proof Statement (2.9) follows from

$$\delta^p(T) = c(\begin{bmatrix} I & D^{-1}N\end{bmatrix}) = c(D) + c(\begin{bmatrix} I & D^{-1}N\end{bmatrix}) = c([D \quad N]).$$

Here the first equality holds because of the above lemma, whereas the second and third equalities hold because of Lemma 2.10 and Lemma 2.11, respectively. To prove (i) we proceed similarly and use the first statement of the previous lemma; we get

$$\delta^p_\alpha(T) = c_\alpha(\begin{bmatrix} I & D^{-1}N\end{bmatrix}) = c_\alpha([D \quad N]) - c_\alpha(D) = -c_\alpha(D),$$

where the last equality holds since $c_\alpha([D \quad N]) = 0$ when $[D(s) \quad N(s)]$ is right $\mathbb{C}^-_\alpha(s)$-unimodular (Lemma 2.12). Because of the nonsingularity of D we have $c_\alpha(D) = c_\alpha(\det\ D)$ (Lemma 2.9) so that (i) follows. Statement (ii) essentially follows from (i): because of (2.2) we have that $\delta^p_\beta(T) = -c_\beta(D)$ for all $\beta \in \mathbb{C}^e \setminus \{\alpha\}$. Since D is nonsingular Lemma 2.10 yields that

$$\sum_{\beta\in\mathbb{C}^e\setminus\{\alpha\}} c_\beta(D) = -c_\alpha(D)$$

so that (ii) follows. ◇

The next theorem considers factorizations of the type $T(s) = N(s)D^{-1}(s)$. The proof of the theorem is completely analogous to the proof of the previous theorem.

Theorem 2.16 Let $T(s) \in \mathbb{C}^{n\times m}(s)$ be given as $T(s) = N(s)D^{-1}(s)$, where $D(s) \in \mathbb{C}^{m\times m}(s)$ and $N(s) \in \mathbb{C}^{n\times m}(s)$. Then

$$\delta^p(T) = c(\begin{bmatrix} N \\ D \end{bmatrix}). \tag{2.10}$$

Moreover, the following implications hold:

(i) $\begin{bmatrix} N(s) \\ D(s) \end{bmatrix}$ is left $\mathbb{C}^-_\alpha(s)$-unimodular $\Longrightarrow$ $\delta^p_\alpha(T) = -c_\alpha(\det\ D)$

(ii) $\begin{bmatrix} N(s) \\ D(s) \end{bmatrix}$ is left $\mathbb{C}^+_\alpha(s)$-unimodular $\Longrightarrow$ $\sum_{\beta\in\mathbb{C}^e\setminus\{\alpha\}} \delta^p_\beta(T) = c_\alpha(\det\ D)$.

2.4 Wiener-Hopf structure of a rational matrix

In contrast with previous sections we make a particular choice of α in this section: we consider $\alpha = \infty$. The material could in fact be presented at a more general level (see Remark 2.26 below), but we prefer to avoid unnecessary abstractness. We will use the notation $\mathbb{C}[s]$ instead of $\mathbb{C}^+_\infty(s)$ and the notation $\mathbb{C}_\infty(s)$ instead of $\mathbb{C}^-_\infty(s)$. Results in this section are given in terms of complex-rational matrices; analogous results for real-rational matrices are obvious.

Let us consider a matrix $T(s) \in \mathbb{C}^{n\times m}(s)$ of full row rank. The following example illustrates that orders of poles at ∞ do not necessarily coincide with row contents at ∞.

Example 2.17 Consider the matrix $T(s)$ given by

$$T(s) = \begin{bmatrix} 1 & s^2 & s \\ 0 & s & 0 \end{bmatrix}.$$

Then

$$T(s) = U(s)D(s)V(s) \text{ with } D(s) = \begin{bmatrix} s^2 & 0 & 0 \\ 0 & 1 & 0 \end{bmatrix}$$

and

$$U(s) = \begin{bmatrix} 1 & 0 \\ s^{-1} & -1 \end{bmatrix}, \quad V(s) = \begin{bmatrix} s^{-2} & 1 & s^{-1} \\ s^{-1} & 0 & 1 \\ 1 & 0 & 0 \end{bmatrix}.$$

Both $U(s)$ and $V(s)$ are $\mathbb{C}_\infty(s)$-unimodular, so $T(s)$ has two poles at ∞ of orders 2 and 0 respectively. The orders of the poles at ∞ do *not* coincide with the row contents at ∞, which are 2 and 1. ◇

There are however situations where the orders of the poles at ∞ *do* coincide with the row contents at ∞. Indeed, it follows from the theorem on the Smith-McMillan form at ∞ (Theorem 2.6) that there exists a $\mathbb{C}_\infty(s)$-unimodular matrix $U_-(s)$ such that the orders of the poles at ∞ of $U_-(s)T(s)$ coincide with the row contents at ∞ of $U_-(s)T(s)$. In fact, the matrix $U_-(s)T(s)$ has a special property that is called "row reduced", a terminology which stems from [56, 61]. The importance of row reducedness for polynomial matrices was recognized earlier in the work of Wolovich [71] who termed such matrices "row proper". The precise definition is as follows:

Definition 2.18 Let $T(s) \in \mathbb{C}^{n\times m}(s)$ have full row rank. Denote the j-th row of $T(s)$ by $t_j(s)$ and let $\mu_j = c_\infty(t_j)$ for $j = 1, \ldots, n$. Let $\hat{T}(s)$ be defined as

$$\hat{T}(s) = \text{diag }(s^{-\mu_1}, \ldots, s^{-\mu_n})T(s).$$

Then $T(s)$ is *row reduced* if $\hat{T}(s)$ is right $\mathbb{C}_\infty(s)$-unimodular, i.e. $\hat{T}(\infty)$ has full row rank. ◇

In the above we saw that $T(s)$ can be brought in row reduced form by left multiplication by a $\mathbb{C}_\infty(s)$-unimodular matrix $U_-(s)$. The next theorem from [71] shows that this can also be accomplished by left multiplication by a $\mathbb{C}[s]$-unimodular matrix $U_+(s)$.

Theorem 2.19 Every matrix $T(s) \in \mathbb{C}^{n\times m}(s)$ of rank r can be factorized as

$$T(s) = U_+(s)D(s)U_-(s) \text{ with } D(s) = \begin{bmatrix} \Delta(s) & 0 \\ 0 & 0 \end{bmatrix} \tag{2.11}$$

where $U_+(s)$ is a $\mathbb{C}[s]$-unimodular matrix of size $n \times n$ and $U_-(s)$ is a $\mathbb{C}_\infty(s)$-unimodular matrix of size $m \times m$; $\Delta(s) = \text{diag }(s^{\kappa_1}, \ldots, s^{\kappa_r})$ with $\kappa_1 \geq \kappa_2 \geq \cdots \geq \kappa_r$. Moreover, the integers κ_i are unique $(i = 1, \ldots, r)$.

Proof Applying Theorem 2.2 for $\mathcal{R} = \mathbb{C}[s]$ we conclude that a $\mathbb{C}[s]$-unimodular matrix $V_+(s)$ can be found such that $V_+(s)T(s)$ has the form

$$\begin{bmatrix} \tilde{T}(s) \\ 0 \end{bmatrix}$$

where $\tilde{T}(s)$ has full row rank. Applying the same theorem for $\mathcal{R} = \mathbb{C}_\infty(s)$, we conclude that a $\mathbb{C}_\infty(s)$-unimodular matrix $V_-(s)$ can be found such that $\tilde{T}(s)V_-(s)$ has the form $[\bar{T}(s) \quad 0]$ where $\bar{T}(s)$ is nonsingular. It is therefore sufficient to prove the theorem for the case that $T(s)$ is invertible, and we may then proceed as in the proof of Theorem 2.5.7 in [71]. Denoting the j-th row of $T(s)$ by $t_j(s)$, we define the integers $\mu_j = c_\infty(t_j)$ for $j = 1, \ldots, r$. Permuting rows if necessary, we may assume that $\mu_1 \geq \mu_2 \geq \cdots \geq \mu_r$. Clearly

$$\mu := \sum_{i=1}^{r} \mu_i \geq c_\infty(T) = c_\infty(\det T). \tag{2.12}$$

Next define

$$\hat{T}(s) := \text{diag }(s^{-\mu_1}, \ldots, s^{-\mu_r})T(s).$$

Clearly $\hat{T}(s)$ is a matrix over $\mathbb{C}_\infty(s)$; if $\hat{T}(\infty)$ is invertible then $\hat{T}(s)$ is $\mathbb{C}_\infty(s)$-unimodular (Lemma 2.12) and we have a factorization of the desired type. So let us assume that $\hat{T}(\infty)$ is not invertible. Let r_j be a row of $\hat{T}(\infty)$ that is dependent on its successor rows:

$$r_j = \sum_{i=j+1}^{r} a_i r_i.$$

Define a constant invertible matrix V as the $r \times r$ identity matrix except for the j-th row which is defined as $[0 \ldots -1 \quad a_{j+1} \ldots a_r]$. Then the j-th row of $V\hat{T}(\infty)$ equals zero so that the content at ∞ of the j-th row of $V\hat{T}(s)$ is negative, while the other rows equal those of $\hat{T}(s)$. Define

$$V(s) := \text{diag }(s^{\mu_1}, \ldots, s^{\mu_r}) V \text{diag }(s^{-\mu_1}, \ldots, s^{-\mu_r}).$$

Then $V(s)$ is a polynomial matrix that has a constant determinant and is therefore $\mathbb{C}[s]$-unimodular. We then have

$$V(s)T(s) = \text{diag }(s^{\mu_1}, \ldots, s^{\mu_r}) V\hat{T}(s),$$

and it can be easily checked that the sum of the row contents at ∞ of $V(s)T(s)$ is less than μ. The process may now be continued for $V(s)T(s)$ and must come to an end because of (2.12) (note that $c_\infty(VT) = c_\infty(T)$). It remains to prove the uniqueness of the indices κ_j $(j = 1, \ldots, r)$. We shall do this by relating the κ_j's to another set of integers that are, by definition, uniquely determined by $T(s)$. Define, for $k \in \mathbb{Z}$,

$$\nu_k(T) := \dim \{f(\infty) \mid f(s) \in \mathbb{C}_\infty(s) \text{ and } s^{-k}T(s)f(s) \in \mathbb{C}[s]\}. \quad (2.13)$$

The set in the right-hand side of (2.13) is clearly invariant under left multiplication of $T(s)$ by $\mathbb{C}[s]$-unimodular matrices and also under right multiplication by $\mathbb{C}_\infty(s)$-unimodular matrices. Therefore, the integers $\nu_k(T)$ $(k \in \mathbb{Z})$ can be computed from the form $D(s)$ in (2.11), leading to

$$\nu_k(T) = \sharp\{j \mid \kappa_j \geq k\} + \dim \ \ker T(s). \quad (2.14)$$

As a result, $\nu_k - \nu_{k+1} = \sharp\{j \mid \kappa_j = k\}$. This shows that the κ_j's can be expressed in terms of the unique ν_k's. Therefore the κ_j's $(j = 1, \ldots, r)$ must also be unique. ◇

We will call the canonical form $D(s)$ in (2.11) the *left Wiener-Hopf form* of $T(s)$; note that this terminology differs from the terminology in [17] where (2.11) is called a "right Wiener-Hopf factorization at infinity". The integers $\kappa_1, \ldots, \kappa_r$ will be called the *left Wiener-Hopf indices* of $T(s)$. The *left Wiener-Hopf order* of $T(s)$, denoted by $\kappa^\ell(T)$, is defined as

$$\kappa^\ell(T) := \sum_{i=1}^{r} \kappa_i.$$

Remark 2.20 In the proof of Theorem 2.19, the uniqueness of the integers κ_j can also be derived in another way. Instead of relating the κ_j's to the integers $\nu_k(T)$, we can relate them to integers $\mu_k(T)$ that are defined as follows (interpret $\mathbb{C}[s]$ as a vector space over $\mathbb{C}$):

$$\mu_k(T) := \dim \{g(s) \in \mathbb{C}[s] \mid g(s) = s^{-k}T(s)f(s) \text{ for some } f(s) \in \mathbb{C}_\infty(s)\}.$$

It is easily checked (see the proof of the previous theorem) that

$$\mu_k = \sum_{\{j \mid \kappa_j \geq k\}} \{\kappa_j - k + 1\}. \tag{2.15}$$

As a result

$$\mu_k - \mu_{k+1} = \sharp\{j \mid \kappa_j \geq k\}$$

so that

$$\begin{aligned} \sharp\{j \mid \kappa_j = k\} &= (\mu_k - \mu_{k+1}) - (\mu_{k+1} - \mu_{k+2}) \\ &= \mu_k + \mu_{k+2} - 2\mu_{k+1}. \end{aligned}$$

Note that it follows from (2.15) that

$$\mu_1(T) = \kappa^\ell(T).$$

From this it can be easily deduced that

$$\kappa^\ell(T) = \dim \{g(s) \in \mathbb{C}[s] \mid g(s) = T(s)f(s) \text{ for some } f(s) \in s^{-1}\mathbb{C}_\infty(s)\}.$$

This is a fact that will be needed in Chapter 5.

Example 2.21 Let us once more consider the matrix $T(s)$ of Example 2.17. We may write $T(s)$ as $T(s) = U_+(s)D(s)U_-(s)$ with

$$D(s) = \begin{bmatrix} s & 0 & 0 \\ 0 & s & 0 \end{bmatrix}$$

and

$$U_+(s) = \begin{bmatrix} -1 & s \\ 0 & 1 \end{bmatrix}, \quad U_-(s) = \begin{bmatrix} -s^{-1} & 0 & -1 \\ 0 & 1 & 0 \\ 1 & 0 & 0 \end{bmatrix}.$$

The matrix $U_+(s)$ is $\mathbb{C}[s]$-unimodular whereas $U_-(s)$ is $\mathbb{C}_\infty(s)$-unimodular. As a result, the two left Wiener-Hopf indices of $T(s)$ are both equal to 1. ◇

Example 2.17 and Example 2.21 show that left Wiener-Hopf indices do not necessarily coincide with orders of poles at ∞. In the examples they do, however, sum up to the same value. This observation is explained by the next lemma, which in some respect is more general than Corollary 6.4 in [32], which is concerned with nonsingular rational matrices.

Lemma 2.22 Let $T(s) \in \mathbb{C}^{n \times m}(s)$ be a matrix of full row rank. Then

$$\kappa^{\ell}(T) = c_{\infty}(T). \tag{2.16}$$

Moreover, the left Wiener-Hopf indices of $T(s)$ coincide with the orders of the poles at ∞ of $T(s)$ if $T(s)$ is row reduced. In that case they are equal to the row contents at ∞ of $T(s)$.

Proof Factorize $T(s)$ as in Theorem 2.19 as

$$T(s) = U_{+}(s)D(s)U_{-}(s) \text{ with } D(s) = [\Delta(s) \quad ,]0$$

where $D(s)$ is the left Wiener-Hopf form of $T(s)$. Let $V_{-}(s)$ be the right $\mathbb{C}_{\infty}(s)$-unimodular matrix that consists of the first n rows of $U_{-}(s)$. Then

$$T(s) = U_{+}(s)\Delta(s)V_{-}(s).$$

According to Lemma 2.11 we have that $c_{\infty}(T) = c_{\infty}(U_{+}) + c_{\infty}(\Delta) + c_{\infty}(V_{-})$. By definition we have that $c_{\infty}(\Delta) = \kappa^{\ell}(T)$, whereas $c_{\infty}(U_{+}) = c_{\infty}(\det U_{+}) = 0$ because of the $\mathbb{C}[s]$-unimodularity of $U_{+}(s)$. Now Lemma 2.12 implies that $c_{\infty}(V_{-}) = 0$ so that (2.16) follows. Let us now assume that $T(s)$ is row reduced, and let us denote the row contents at ∞ of $T(s)$ by μ_i $(i = 1, \ldots, n)$. Let $\hat{T}(s)$ be defined as in Definition 2.18. Then

$$T(s) = \operatorname{diag}\,(s^{\mu_1}, \ldots, s^{\mu_n})\hat{T}(s)$$

is not only a factorization in left Wiener-Hopf form (2.11) but also a factorization in Smith-McMillan form at ∞ (2.3). This implies the other statements in the lemma. ◇

For a matrix $T(s) \in \mathbb{C}^{n \times m}(s)$ we get the *right Wiener-Hopf form* by interchanging $+$ and $-$ in Theorem 2.19; the *right Wiener-Hopf indices* of $T(s)$ and the *right Wiener-Hopf order* of $T(s)$, denoted by $\kappa^{r}(T)$, are defined in the obvious way.

Theorem 2.23 Every matrix $T(s) \in \mathbb{C}^{n \times m}(s)$ of rank r can be factorized as

$$T(s) = U_{-}(s)D(s)U_{+}(s) \text{ with } D(s) = \begin{bmatrix} \Delta(s) & 0 \\ 0 & 0 \end{bmatrix}$$

where $U_{-}(s)$ is a $\mathbb{C}_{\infty}(s)$-unimodular matrix of size $n \times n$ and $U_{+}(s)$ is a $\mathbb{C}[s]$-unimodular matrix of size $m \times m$; $\Delta(s) = \operatorname{diag}\,(s^{\kappa_1}, \ldots, s^{\kappa_r})$ with $\kappa_1 \geq \kappa_2 \geq \cdots \geq \kappa_r$. Moreover, the integers κ_i are unique $(i = 1, \ldots, r)$.

Proof Apply Theorem 2.19 to the matrix $T^T(s)$. $\diamond$

The following definition is the dual version of Definition 2.18.

Definition 2.24 Let $T(s) \in \mathbb{C}^{n \times m}(s)$ have full column rank. Denote the j-th column of $T(s)$ by $t_j(s)$, and let $\mu_j = c_\infty(t_j)$ for $j = 1, \ldots, m$. Let $\hat{T}(s)$ be defined as

$$\hat{T}(s) = T(s)\text{diag }(s^{-\mu_1}, \ldots, s^{-\mu_m}).$$

Then $T(s)$ is *column reduced* if $\hat{T}(\infty)$ is left $\mathbb{C}_\infty(s)$-unimodular, i.e., $\hat{T}(\infty)$ has full column rank. $\diamond$

Note that both right and left Wiener-Hopf indices are nonnegative when the matrix is polynomial. The reason for this is that a polynomial matrix can be made column (row) reduced by multiplication by a polynomial matrix (Theorem 2.19 and Theorem 2.23).

The next lemma is the analog of Lemma 2.22; its proof is completely analogous and will therefore be omitted.

Lemma 2.25 Let $T(s) \in \mathbb{C}^{n \times m}(s)$ be a matrix of full column rank. Then

$$\kappa^r(T) = c_\infty(T). \tag{2.17}$$

Moreover, the right Wiener-Hopf indices of $T(s)$ coincide with the orders of the poles at ∞ of $T(s)$ if $T(s)$ is column reduced. In that case they are equal to the column contents at ∞ of $T(s)$.

Remark 2.26 The Wiener-Hopf factorization can be defined more generally with respect to a nonempty subset Γ of $\mathbb{C}^e$ for which $\mathbb{C}^e \setminus \Gamma$ is also nonempty. More precisely, there exists an analog of Theorem 2.19 (and of Theorem 2.23) according to which a rational matrix $T(s)$ of rank r can be factorized in the form (2.11), where $U_+(s)$ is now a $\mathbb{C}_\Gamma^+(s)$-unimodular matrix, whereas $U_-(s)$ is a $\mathbb{C}_\Gamma^-(s)$-unimodular matrix. Here $\mathbb{C}_\Gamma^-(s)$ denotes the euclidean domain of rational functions that have no poles in Γ; $\mathbb{C}_\Gamma^+(s)$ denotes the domain of rational functions that have no poles in $\mathbb{C}^e \setminus \Gamma$. The matrix $\Delta(s)$ in the factorization is given by $\Delta(s) = \text{diag }(\chi^{\kappa_1}(s), \ldots, \chi^{\kappa_r}(s))$ where $\chi(s)$ is a rational function that has exactly one pole in Γ and one zero in $\mathbb{C}^e \setminus \Gamma$. Apart from the case $\Gamma = \{\infty\}$ of this subsection, we might also consider the following cases:

(i) $\Gamma = \{s \in \mathbb{C} \mid \ |s| \geq 1\}$

(ii) $\Gamma = \{s \in \mathbb{C} \mid \ \text{Re } s \geq 0\}$.

Case (i) corresponds to a factorization with respect to what is called a "standard contour" in operator theory. The generalized version of Lemma 2.22 yields that the Wiener-Hopf order with respect to Γ of a rational matrix $T(s)$ equals the number of poles of $T(s)$ in Γ minus the number of zeros of $T(s)$ in Γ. Consequently, the order equals the "winding number" of the determinant of $T(s)$ if $T(s)$ is nonsingular (and has no poles or zeros on the contour). It may come as no surprise that the Wiener-Hopf factorization originated from the field of operator theory. A standard reference on the subject is [9].

2.5 Minimal basis of a rational vector space

In the previous sections we were concerned with structural properties of rational matrices. In this section we shall explore the structure of rational *vector spaces.* As before we use the notation $\mathbb{C}[s]$ for the ring of polynomials and the notation $\mathbb{C}_\infty(s)$ for the ring of proper rational functions; again results are formulated over $\mathbb{C}$ but can be reformulated over $\mathbb{R}$ in the obvious way. Recall that the content at infinity $c_\infty(p)$ of a polynomial vector $p(s) \in \mathbb{C}^n[s]$ is defined as the maximum degree of the components of $p(s)$.

The following definition repeats the concept of a "minimal polynomial basis" as given by Forney in [15]; the terminology stems from [56]. In the definition below "col$(g_1, g_2, \ldots, g_r)$" denotes a matrix with columns $g_1, g_2, \ldots, g_r$.

Definition 2.27 Let $\mathcal{X}$ be an r-dimensional subspace of $\mathbb{C}^n(s)$. Let $\{g_1, g_2, \ldots, g_r\}$ be a basis for $\mathcal{X}$ with $g_i \in \mathbb{C}^n[s]$ for $i = 1, \ldots, r$ and $c_\infty(g_1) \geq c_\infty(g_2) \geq \cdots \geq c_\infty(g_r)$. Define $T(s) \in \mathbb{C}^{n \times r}(s)$ by $T(s) := \text{col}(g_1, g_2, \ldots, g_r)$. Then $\{g_1, g_2, \ldots, g_r\}$ is called a *minimal polynomial basis* for $\mathcal{X}$ if $T(s)$ is left $\mathbb{C}[s]$-unimodular and column reduced.

For $m \neq 0$ the *minimal indices* of $\mathcal{X}$, denoted by $\nu_i(\mathcal{X})$ $(i = 1, \ldots, r)$, are defined as

$$\nu_i(\mathcal{X}) = c_\infty(g_i).$$

The *minimal order* of $\mathcal{X}$, denoted by $\nu(\mathcal{X})$, is defined as

$$\nu(\mathcal{X}) = \sum_{i=1}^{r} \nu_i(\mathcal{X}).$$

For $r = 0$ the value of $\nu(\mathcal{X})$ is defined to be zero. ◇

Note that $\nu(\mathbb{C}^n(s)) = 0$ since the set of unit vectors is a minimal polynomial basis for $\mathbb{C}^n(s)$.

It is not difficult to see that, for a given subspace $\mathcal{X}$, there always exists a minimal polynomial basis. The reason for this is that any matrix $T_1(s) \in \mathbb{C}^{n\times r}(s)$ of full column rank can be factorized as

$$T_1(s) = T(s)T_2(s) \tag{2.18}$$

where $T(s)$ is left $\mathbb{C}[s]$-unimodular and column reduced and $T_2(s)$ is nonsingular. This follows from Theorem 2.2 (take $\mathcal{R} = \mathbb{C}[s]$) and Theorem 2.23. The ν_i's are well-defined because of the fact that left $\mathbb{C}[s]$-unimodular matrices $T(s)$ and $\tilde{T}(s)$ for which $\text{im } T(s) = \text{im } \tilde{T}(s)$ are necessarily related via $T(s) = \tilde{T}(s)F(s)$, where $F(s)$ is a $\mathbb{C}[s]$-unimodular matrix. As a result, the right Wiener-Hopf indices of $T(s)$ and $\tilde{T}(s)$ are the same; if both $T(s)$ and $\tilde{T}(s)$ are column reduced, the column degrees of the matrices coincide (Lemma 2.25). The next lemma clarifies the relation between the Wiener-Hopf indices and the content of a rational matrix on the one hand and the minimal indices of related rational vector spaces on the other hand. Part of the results in the lemma can be found in [56, 61].

Lemma 2.28 Let $T(s) \in \mathbb{C}^{n\times m}(s)$. Then

$$c(T) = \nu(\text{ im } T(s)) + \nu(\text{ ker } T(s)). \tag{2.19}$$

Moreover, if $T(s)$ is a polynomial matrix, then the following statements hold:

(i) $\nu(\text{ im } T(s)) \leq \kappa^r(T)$ with equality if $T(s)$ is left $\mathbb{C}[s]$-unimodular; in that case the right Wiener-Hopf indices of $T(s)$ coincide with the minimal indices of $\text{im } T(s)$

(ii) $\nu(\text{ ker } T(s)) \leq \kappa^\ell(T)$ with equality if $T(s)$ is right $\mathbb{C}[s]$-unimodular; in that case the left Wiener-Hopf indices of $T(s)$ coincide with the minimal indices of $\text{im } T^T(s)$.

Proof Let us first assume that $T(s)$ is left $\mathbb{C}[s]$-unimodular. According to Theorem 2.23 there exists a $\mathbb{C}[s]$-unimodular matrix $U_+(s)$ such that $T(s)U_+(s)$ is column reduced. The columns of $T(s)U_+(s)$ are then a minimal polynomial basis for im $T(s)$; the minimal indices of im $T(s)$ are the column degrees of $T(s)U_+(s)$, which clearly coincide with the right Wiener-Hopf indices of $T(s)$. In particular we then have

$$\nu(\text{ im } T(s)) = \kappa^r(T) = c_\infty(T) = c(T), \tag{2.20}$$

where the last two equalities hold because of Lemma 2.25 and Lemma 2.13, respectively. This proves the last two statements in (i) as well as the last statement in (ii). Let us next assume that $T(s)$ is a rational matrix of full column rank. Factorize $T(s)$ as in Theorem 2.6 as

$$T(s) = U(s)D(s)V(s) \text{ with } D(s) = \begin{bmatrix} \Delta(s) \\ 0 \end{bmatrix}$$

where $D(s)$ is the Smith-McMillan form at ∞ of $T(s)$. Let $U_1(s)$ be the left $\mathbb{C}[s]$-unimodular matrix that consists of the first m columns of $U(s)$. Then

$$T(s) = U_1(s)\Delta(s)V(s)$$

so that $c(T) = c(U_1) + c(\Delta) + c(V) = c(U_1)$ (use Lemma 2.11 and Lemma 2.10). It follows from (2.20) that $c(U_1) = \nu(\text{ im } U_1(s))$. Since $\text{im } T(s) = \text{ im } U_1(s)$, we have

$$c(T) = \nu(\text{ im } T(s)), \tag{2.21}$$

and this implies that (2.19) holds for matrices of full column rank.

Let us next assume that $T(s)$ has full row rank. Factorize $T(s)$ as $T(s) = T_1(s)\,[I_{n\times n} \quad T_2(s)]$ where $T_1(s)$ is nonsingular. Setting $\tilde{n} := m-n$ we then have that

$$\ker T(s) = \text{ im } \begin{bmatrix} T_2(s) \\ -I_{\tilde{n}\times\tilde{n}} \end{bmatrix}$$

and

$$c(T) = c(T_1) + c([I_{n\times n} \quad T_2(s)]) = c([I_{n\times n} \quad T_2(s)]) = c(\begin{bmatrix} T_2(s) \\ -I_{\tilde{n}\times\tilde{n}} \end{bmatrix})$$

(use Lemma 2.11, Lemma 2.10 and Lemma 2.14). Using the above proven fact that (2.21) holds for matrices of full column rank, we conclude that

$$c(T) = c(\begin{bmatrix} T_2(s) \\ -I_{\tilde{n}\times\tilde{n}} \end{bmatrix}) = \nu(\begin{bmatrix} T_2(s) \\ -I_{\tilde{n}\times\tilde{n}} \end{bmatrix}) = \nu(\text{ ker } T(s)), \tag{2.22}$$

and this implies that (2.19) holds for matrices of full row rank. But then (2.19) also holds for general matrices $T(s)$, since we may factorize $T(s)$ as $T(s) = T_1(s)T_2(s)$ where $T_1(s)$ has full column rank and $T_2(s)$ has full row rank. It follows from Lemma 2.11 and the above that

$$c(T) = c(T_1) + c(T_2) = \nu(\text{ im } T_1(s)) + \nu(\text{ ker } T_2(s)).$$

Since $\text{im } T(s) = \text{ im } T_1(s)$ and $\ker T(s) = \ker T_2(s)$, this implies (2.19).

Let us now assume that $T(s)$ is a polynomial matrix and let us denote the rank of $T(s)$ by r. Factorize $T(s)$ as in Theorem 2.19 as

$$T(s) = U_+(s)D(s)U_-(s) \text{ with } D(s) = \begin{bmatrix} \Delta(s) & 0 \\ 0 & 0 \end{bmatrix}$$

where $D(s)$ is the left Wiener-Hopf form of $T(s)$. Let $V_+(s)$ be the left $\mathbb{C}[s]$-unimodular matrix that consists of the first r columns of $U_+(s)$. Let $V_-(s)$ be the right $\mathbb{C}_\infty(s)$-unimodular matrix that consists of the first r rows of $U_-(s)$. Then

$$T(s) = V_+(s)\Delta(s)V_-(s).$$

According to Lemma 2.11 we have that $c_\infty(T) = c_\infty(V_+)+c_\infty(\Delta)+c_\infty(V_-)$. By definition $c_\infty(\Delta) = \kappa^\ell(T)$, whereas $c_\infty(V_-) = 0$ by Lemma 2.12. Since $\text{im } V_+(s) = \text{im } T(s)$, we conclude from (2.20) that $\nu(\text{ im } T(s)) = c_\infty(V_+) = c_\infty(T) - \kappa^\ell(T)$. Using (2.19) we get

$$\begin{aligned}\nu(\text{ ker } T(s)) &= c(T) - \nu(\text{ im } T(s)) \\ &= c(T) - (c_\infty(T) - \kappa^\ell(T)) \\ &= \kappa^\ell(T) + \sum_{\beta\in\mathbb{C}} c_\beta(T) \\ &\leq \kappa^\ell(T).\end{aligned}$$

Clearly, equality holds if $T(s)$ is right $\mathbb{C}[s]$-unimodular (Lemma 2.13), and this proves (ii). The inequality in (i) is proven in a completely similar way.

2.6 Preliminary results for matrix pencils

In this section we consider matrices of the form $sE - A$ where E and A are constant real-valued matrices of size $m \times n$. A polynomial matrix of this form is generally called a "matrix pencil". The terminology originates from geometry where a one-parameter family of curves is referred to as a "pencil" of curves. We interpret E and A as linear mappings from X to Z where $X = \mathbb{R}^n$ and $Z = \mathbb{R}^m$. In the sequel the space of rational X-valued functions in s is denoted by $X(s)$; the notation $X_\infty(s)$ is used for the space of proper rational X-valued functions. Writing the Laurent expansion of an element $f(s) \in X(s)$ as

$$f(s) = f_{-\ell}s^\ell + f_{-\ell+1}s^{\ell-1} + \cdots + f_0 + f_1 s^{-1} + f_2 s^{-2} + \cdots,$$

we define the projection $\pi_1 \colon X(s) \to \mathbb{R}$ by

$$\pi_1 : f(s) \mapsto f_1. \tag{2.23}$$

Let us introduce the following iteration

$$V^0 = X, \quad V^{m+1} = A^{-1}EV^m \tag{2.24}$$

where by standard notation $A^{-1}V := \{x \mid Ax \in V\}$. Clearly

$$V^0 \supset V^1 \supset V^2 \supset \cdots$$

and a limit space V^* is reached in finitely many steps. The space V^* is the largest subspace $V \subset X$ for which $AV \subset EV$.

Lemma 2.29 Let $E, A : X \to Z$ be linear mappings. Let V^* be the limit space of the iteration (2.24). Then $x \in V^*$ if and only if there exists $x(s) \in s^{-1}X_\infty(s)$ such that

$$\pi_1 x(s) = x \quad \text{and} \quad (sE - A)x(s) = Ex.$$

Proof (cf. [23]) The "if" part is proven as follows. Writing the Laurent expansion of $x(s)$ as

$$x(s) = x_1 s^{-1} + x_2 s^{-2} + \cdots$$

we have that $x_1 = x$ and $Ex_{i+1} = Ax_i$ for all $i \geq 1$. We conclude that $x \in V^k$ for all $k \geq 0$ so that $x \in V^*$. To prove the "only if" part let us choose bases in X and Z such that E is given as

$$E = \begin{bmatrix} I & 0 \\ 0 & 0 \end{bmatrix}.$$

Accordingly, we write

$$A = \begin{bmatrix} A_{11} & A_{12} \\ A_{21} & A_{22} \end{bmatrix}.$$

Let $x \in V^*$ be written accordingly as

$$x = \begin{bmatrix} x_1 \\ x_2 \end{bmatrix} \tag{2.25}$$

and let $V_1^* := \pi_{X_1} V^*$ where π_{X_1} denotes projection onto X_1 along X_2. Then V_1^* is the largest subspace $V_1 \subset X_1$ for which

$$\begin{bmatrix} A_{11} \\ A_{21} \end{bmatrix} V_1 \subset V_1 \times \{0\} + \text{ im } \begin{bmatrix} A_{12} \\ A_{22} \end{bmatrix}.$$

It is easily checked that there exists a linear mapping $F : X_1 \to X_2$ such that not only

$$(A_{11} + A_{12}F)V_1^* \subset V_1^* \text{ and } (A_{21} + A_{22}F)V_1^* = \{0\} \tag{2.26}$$

but also $Fx_1 = x_2$ (use (2.25)). Defining $x_1(s) = (sI - A_{11} - A_{12}F)^{-1} x_1$ and $x_2(s) = Fx_1(s)$ we have that

$$\begin{bmatrix} sI - A_{11} & -A_{12} \\ -A_{21} & -A_{22} \end{bmatrix} \begin{bmatrix} x_1(s) \\ x_2(s) \end{bmatrix} = \begin{bmatrix} x_1 \\ 0 \end{bmatrix} = Ex$$

which proves the lemma.

Remark 2.30 Because of the fact that V^* is a subspace for which $AV^* \subset EV^*$ we also have

$$x \in V^* \implies \exists\, x(s) \in s^{-1}X_\infty(s) \text{ such that } (sE - A)x(s) = Ax.$$

Theorem 2.31 ([24]) Let $E, A : X \to Z$ be linear mappings. Let V^* be the limit space of the iteration (2.24). Then $V^* = \{0\}$ if and only if the matrix $sE - A$ has full column rank for all $s \in \mathbb{C}$.

Proof To prove the "if" part let us assume that $sE - A$ has full column rank for all $s \in \mathbb{C}$ and that $x \in V^*$. Then there exists a polynomial matrix $L(s)$ such that $L(s)(sE - A) = I$ (Lemma 2.13). By the previous lemma there exists $x(s) \in s^{-1}X_\infty(s)$ with $\pi_1 x(s) = x$ such that $(sE - A)x(s) = Ex$. Then $x(s) = L(s)Ex$ is polynomial so that necessarily $x(s) = 0$, and in particular $x = 0$. Conversely, let us assume that $V^* = \{0\}$. Let $s_0 \in \mathbb{C}$ be arbitrary and let $x \in X$ be such that $(s_0 E - A)x = 0$. Define

$$\tilde{x}(s) := \frac{1}{s - s_0} x.$$

Then $\tilde{x}(s) \in s^{-1}\mathbb{C}^n_\infty(s)$, $\pi_1 \tilde{x}(s) = x$ and

$$\begin{aligned}(sE - A)\tilde{x}(s) &= (sE - A)\tilde{x}(s) - (s_0 E - A)\tilde{x}(s) \\ &= (s - s_0)E\tilde{x}(s) \\ &= Ex.\end{aligned}$$

Let us now define $x(s)$ as the element of $s^{-1}X_\infty(s)$ that is obtained by projecting all coefficients of $\tilde{x}(s)$ on their real parts. Clearly $\pi_1 x(s) = x$ and $(sE - A)x(s) = Ex$, so that it follows from the previous lemma that $x \in V^*$, which implies that $x = 0$. We conclude that $s_0 E - A$ has full column rank. ◇

The iteration (2.24) is related to a number of characteristics of the pencil $sE - A$. It has been shown in [6, Theorem 2] and [4, Theorem 4] that

$$\dim\, (V^* \cap \ker E) = \dim\, \ker\, (sE - A). \tag{2.27}$$

We will give a different proof of this result in Theorem 6.38, where we show as well that the subspaces $V^k \cap \ker E$ determine the zero structure at infinity of $sE - A$.

Decomposing X as $X = X_1 \oplus X_2$ with $X_1 = V^*$ and decomposing Z as $Z = Z_1 \oplus Z_2$ with $Z_1 = EV^*$, we may write $sE - A$ as (note that $AV^* \subset EV^*$)

$$\begin{bmatrix} sE_{11} - A_{11} & sE_{12} - A_{12} \\ 0 & sE_{22} - A_{22} \end{bmatrix}. \tag{2.28}$$

It follows from Theorem 2.31 that $sE_{22} - A_{22}$ has full column rank for all $s \in \mathbb{C}$ whereas, by construction, E_{11} is a matrix of full row rank. In fact, it follows from the Kronecker canonical form (see, for instance, [19, Chapter XII-4]) that bases in X and Z can be found such that with respect to these bases the matrix $sE - A$ is given in the form (2.28) with $E_{12} = 0$ and $A_{12} = 0$. In the Kronecker canonical form the submatrix $sE_{11} - A_{11}$ has the form diag $\{(sE_i - A_i)\}_i$ where the nonzero blocks $(sE_i - A_i)$ (which may have different sizes) can be of

$$\text{Type I:} \quad E_i = \begin{bmatrix} 1 & 0 & 0 \\ 0 & 1 & 0 \\ 0 & 0 & 1 \end{bmatrix}, A_i = \begin{bmatrix} \mu_i & 1 & 0 \\ 0 & \mu_i & 1 \\ 0 & 0 & \mu_i \end{bmatrix} \qquad \mu_i \in \mathbb{C}$$

and/or of

$$\text{Type II:} \quad E_i = \begin{bmatrix} 1 & 0 & 0 \\ 0 & 1 & 0 \end{bmatrix}, A_i = \begin{bmatrix} 0 & 1 & 0 \\ 0 & 0 & 1 \end{bmatrix}.$$

The submatrix $sE_{22} - A_{22}$ comprises blocks of

$$\text{Type III:} \quad E_i = \begin{bmatrix} 0 & 1 & 0 \\ 0 & 0 & 1 \\ 0 & 0 & 0 \end{bmatrix}, A_i = \begin{bmatrix} 1 & 0 & 0 \\ 0 & 1 & 0 \\ 0 & 0 & 1 \end{bmatrix}$$

and/or of

$$\text{Type IV:} \quad E_i = \begin{bmatrix} 1 & 0 \\ 0 & 1 \\ 0 & 0 \end{bmatrix}, A_i = \begin{bmatrix} 0 & 0 \\ 1 & 0 \\ 0 & 1 \end{bmatrix}.$$

Blocks of type I completely determine the *finite zero structure* of $sE - A$: a zero at $\mu_i \in \mathbb{C}$ of order $k \geq 1$ gives rise to a block of type I of size $k \times k$; in the terminology of Kronecker, k is the degree of a "finite elementary divisor" of $sE - A$. Blocks of type III completely determine the *infinite zero structure* of $sE - A$: a zero at ∞ of order $k \geq 1$ gives rise to a block of type III of size $(k+1) \times (k+1)$; in the terminology of Kronecker, $k+1$ is the degree of an "infinite elementary divisor" of $sE - A$.

Blocks of type II are not present if the matrix $sE - A$ has full column rank. Blocks of this type correspond to the nonzero minimal indices (see Definition 2.27) of the rational vector space ker $(sE - A)$: a minimal index with value $k \geq 1$ gives rise to a block of type II of size $k \times (k+1)$. The indices are commonly referred to as the *right Kronecker indices* of $sE - A$. In an analogous way blocks of type IV correspond to the nonzero minimal indices of the left null space of $sE - A$, i.e., to the nonzero minimal indices of ker $(sE - A)^T$. These indices are commonly referred to as the *left Kronecker indices* of $sE - A$. In Theorem 6.2 we will relate both the left and right Kronecker indices to certain subspace recursions involving E and A.

Remark 2.32 The right Kronecker indices of matrix pencils of the form $[sI - A \quad B]$ coincide with the "controllability indices" related to the pair (A, B) as classically defined in [71, 72]. A proof of this fact is given in [51, Theorem 3.1.1] under the extra condition that the matrix $[sI - A \quad B]$ has full row rank for all $s \in \mathbb{C}$. This condition is however inessential in the present context. By duality, the left Kronecker indices of matrix pencils of the form

$$\begin{bmatrix} sI - A \\ C \end{bmatrix}$$

coincide with the classically defined "observability indices". ◇

Next, we define a different type of iteration that plays a dual role as compared to the iteration (2.24). The iteration is defined by

$$T^0 = \{0\}, \quad T^{m+1} = E^{-1}AT^m. \tag{2.29}$$

Clearly

$$T^0 \subset T^1 \subset T^2 \subset \cdots$$

and a limit space T^* is reached in finitely many steps. The space T^* is the smallest subspace $V \subset Z$ with $V \supset E^{-1}AV$. The following theorem is the analog of Theorem 2.31.

Theorem 2.33 Let $E, A : X \to Z$ be linear mappings. Let T^* be the limit space of the iteration (2.29). Then $AT^* = Z$ if and only if $sE - A$ has full row rank for all $s \in \mathbb{C}$.

Proof Note that $sE - A$ has full row rank for all $s \in \mathbb{C}$ if and only if $sE^T - A^T$ has full column rank for all $s \in \mathbb{C}$. Let us now introduce the following iteration in Z:

$$S^0 = \{0\}, \quad S^{m+1} = AE^{-1}S^m. \tag{2.30}$$

Clearly, $AT^k = S^k$ for $k \geq 0$. Using some elementary results from linear algebra we conclude from Theorem 2.31 that $sE^T - A^T$ has full column rank for all $s \in \mathbb{C}$ if and only if $S^* = Z$, which is, of course, equivalent to $AT^* = Z$. ◇

It will be shown in Theorem 6.38 that

$$\dim\,(AT^* + \text{ im } E) = \text{rank}\,(sE - A)$$

and that the subspaces $AT^k +$ im E determine the zero structure at infinity of $sE - A$ (note the duality with (2.27)).

We conclude this section with a number of lemmas that are corollaries of earlier results and will be needed in subsequent chapters. The first lemma presents a well-known result (see, for instance, [24]).

Lemma 2.34 Let A and C be constant matrices of sizes $n \times n$ and $p \times n$, respectively. Then the following statements are equivalent:

(i) $\begin{bmatrix} sI - A \\ C \end{bmatrix}$ has full column rank for all $s \in \mathbb{C}$

(ii) $C(sI - A)^{-1}x = 0 \implies x = 0$.

Proof Let V^* be the limit space of the iteration (2.24) applied to the pencil

$$\begin{bmatrix} sI - A \\ C \end{bmatrix}.$$

According to Theorem 2.31 statement (i) is equivalent to

$$V^* = \{0\}. \tag{2.31}$$

On the other hand, writing the equation $C(sI - A)^{-1}x = 0$ as

$$\begin{bmatrix} sI - A \\ C \end{bmatrix} (sI - A)^{-1}x = \begin{bmatrix} I \\ 0 \end{bmatrix} x,$$

we conclude from Lemma 2.29 that (ii) is also equivalent to (2.31).

Lemma 2.35 Let $E, A : X \to Z$ and $C : X \to Y$ be linear mappings. Let V^* and S^* be limit spaces of iteration (2.24) applied to the pencils $sE - A$ and

$$\begin{bmatrix} sE - A \\ C \end{bmatrix}$$

respectively:

$$V^0 = X, \quad V^{m+1} = A^{-1}EV^m$$

$$S^0 = X, \quad S^{m+1} = A^{-1}ES^m \cap \ker C.$$

Let $\bar{S}^*$ be the limit space of the iteration

$$\bar{S}^0 = V^*, \quad \bar{S}^{m+1} = A^{-1}E\bar{S}^m \cap \ker C.$$

Then $\bar{S}^* = S^*$.

Proof It follows from the definitions that $\bar{S}^k \subset S^k \subset V^k$ for all $k \geq 0$. This implies that

$$\bar{S}^* \subset S^* \tag{2.32}$$

and $S^* \subset V^*$. From the last inclusion it follows that

$$S^* = A^{-1}ES^* \cap \ker C \subset A^{-1}EV^* \cap \ker C = \bar{S}^1.$$

By induction we conclude that $S^* \subset \bar{S}^k$ for all $k \geq 0$ so that

$$S^* \subset \bar{S}^*.$$

Together with (2.32) this gives the desired result. ◇

The next two lemmas generalize Theorem 10 and Theorem 8 in [4], respectively.

Lemma 2.36 Let $E, A : X \to Z$ be linear mappings and let T^* be the limit space of (2.29). Then the following statements are equivalent:

(i) $A^{-1}[\operatorname{im} E] \cap \ker E = \{0\}$

(ii) $sE - A$ has full column rank and $\delta_\infty^z(sE - A) = 0$.

Moreover, if one of the above statements hold then $T^* = \ker E$.

Proof Let us first assume that (i) holds and prove that $T^* = \ker E$. By definition $T^2 = E^{-1}AT^1 = E^{-1}A[\ker E]$, and this is easily seen to be equal to $\ker E$. By induction it follows that $T^k = \ker E$ for all $k \in \mathbb{Z}_+$ so that $T^* = \ker E$. Next, we prove that (ii) holds. For any $f(s) \in X_\infty(s)$ with Laurent expansion

$$f(s) = f_0 + f_1 s^{-1} + \cdots$$

we have that $(sE - A)f(s) = (Ef_0)s + Ef_1 - Af_0 + \cdots$. Clearly $(sE - A)f(s) = 0$ implies that $f_i \in A^{-1}[\operatorname{im} E] \cap \ker E = \{0\}$ for all $i \in \mathbb{Z}_+$. This proves that $sE - A$ has full column rank. Next, if $(sE - A)f(s)$ is strictly proper then $f(s)$ itself has to be strictly proper since in that case $f_0 \in A^{-1}[\operatorname{im} E] \cap \ker E = \{0\}$. It now follows from Lemma 2.8 that, in the terminology of the lemma, $s_{-1} = 0$, i.e., $\delta_\infty^z(sE - A) = 0$. This proves that (i) implies (ii). Let us next assume that (ii) holds. Let $f_0 \in A^{-1}[\operatorname{im} E] \cap \ker E$; let f_1 be such that $Ef_1 = Af_0$. Defining $f(s) = f_0 + f_1 s^{-1}$ we have that $f(s) \in X_\infty(s)$ and $(sE - A)f(s) = (-Af_1)s^{-1}$. Now Lemma 2.8 implies that either $f_0 = 0$, which ends the proof, or $Af_1 = 0$. If $Af_1 = 0$ it follows from the fact that $sE - A$ has full column rank that $f(s) = 0$ and in particular that $f_0 = 0$. We conclude that (i) holds. ◇

By the duality argument of the proof of Theorem 2.33, the following lemma is immediate from the previous one.

Lemma 2.37 Let $E, A : X \to Z$ be linear mappings and let V^* be the limit space of (2.24). Then the following statements are equivalent:

(i) $A[\ker E] + \operatorname{im} E = Z$

(ii) $sE - A$ has full row rank and $\delta_\infty^z(sE - A) = 0$.

Moreover, if one of the above statements holds, then $EV^* = \operatorname{im} E$.

Lemma 2.38 Let E and A be constant matrices. Then

$$\operatorname{rank} E = \delta^p(sE - A).$$

Proof For all $f(s) \in X_\infty(s)$ the vector $s^{-2}(sE-A)f(s)$ is strictly proper. It therefore follows from Lemma 2.8 that $sE-A$ does not have poles at ∞ of order ≥ 2. Again using Lemma 2.8 we then have that $\delta^p_\infty(sE-A)$ is given by

$$\dim \{g(\infty) \mid g(s) \in Z_\infty(s), g(s) = (sE-A)f(s) \quad \text{for some } f(s) \in s^{-1}X_\infty(s)\}$$

which is easily seen to be identical to rank E. Since $\delta^p_\beta(sE-A) = 0$ for all $\beta \in \mathbb{C}$ it follows that

$$\delta^p_\infty(sE-A) = \delta^p(sE-A)$$

and this proves the lemma. ◇

The next lemma will be very useful for proving minimality results in Chapter 4; the lemma connects the rank of a constant matrix E to properties of the matrix $sE-A$ and related rational vector spaces.

Lemma 2.39 Let E and A be constant matrices. Then

$$\text{rank } E = \delta^z(sE-A) + \nu(\text{ ker } (sE-A)) + \nu(\text{ im } (sE-A)).$$

Proof According to the previous lemma we have rank $E = \delta^p(sE-A)$, which is by definition equal to $c(sE-A)+\delta^z(sE-A)$. Applying Lemma 2.28 yields the desired result.

3

Representations of linear time-invariant systems

As already mentioned in Chapter 1, we adopt the approach in [64]–[68] in which a fundamental distinction is made between dynamical systems and their mathematical representations. In the next section we repeat the formal definition of a dynamical system as given by Willems [66]. In subsequent sections we consider various types of representations of dynamical systems. Polynomial matrices will play a fundamental role throughout the chapter. In fact, the study of behaviors in this book relies essentially on a module structure. Details are provided by the next section, which is tutorial and partly in the spirit of [7, Chapter 5]; we do not pretend the results to be new.

3.1 Dynamical systems

Definition 3.1 A *dynamical system* Σ is defined as a triple $\Sigma := (\mathbf{T}, W, \mathcal{B})$ where $\mathbf{T} \subseteq \mathbb{R}$ is the *time set*, W is the *space of external variables*, and $\mathcal{B}$ is a subset of $W^{\mathbf{T}}$ called the *behavior* of the system. ◇

In this book we take $W = \mathbb{R}^q$. We are aiming at a unified treatment of the discrete-time case and the continuous-time case; we will consider the case $\mathbf{T} = \mathbb{Z}_+$ and the case $\mathbf{T} = \mathbb{R}_+ = (0, \infty)$ (results are also valid for $\mathbf{T} = \mathbb{R}$). We will formulate our results in terms of a linear mapping σ from a linear space $\mathcal{D} \subset \mathbb{R}^{\mathbf{T}}$ to itself; we will assume that $\mathcal{B} \subset \mathcal{D}^q$. In the case $\mathbf{T} = \mathbb{Z}_+$ we simply take $\mathcal{D} = \mathbb{R}^{\mathbb{Z}_+}$ and the mapping σ is defined as "shift":

$$\sigma((w_0, w_1, w_2, \ldots)) := (w_1, w_2, \ldots).$$

In the case $\mathbf{T} = \mathbb{R}_+$ the mapping σ is defined as "differentiation":

$$\sigma w(t) := \frac{dw}{dt}(t), \tag{3.1}$$

and we require $\mathcal{D}$ to be a subspace of $\mathbb{R}^{\mathbb{R}_+}$ that is closed under differentiation; we will specify $\mathcal{D}$ in a moment. When we associate a mapping of the form $p_0+p_1\sigma+\cdots+p_k\sigma^k$ ($p_i \in \mathbb{R}$) to the polynomial $p_0+p_1s+\cdots+p_ks^k \in \mathbb{R}[s]$, then the operations of addition and multiplication are carried over in a nice way. For the case $\mathcal{D} = \mathbb{R}^{\mathbb{Z}_+}$ we have the following properties:

Property 3.1 $p(\sigma)\boldsymbol{x} = 0$ for all $\boldsymbol{x} \in \mathcal{D} \iff p(s) = 0$

Property 3.2 if $p(s) \neq 0$ then

$$\boldsymbol{x}_1 \in \mathcal{D} \implies \text{there exists } \boldsymbol{x}_2 \in \mathcal{D} \text{ with } p(\sigma)\boldsymbol{x}_2 = \boldsymbol{x}_1$$

Property 3.3 $p(s) \neq 0 \implies \dim \ker p(\sigma) = \deg p(s)$.

These properties are essential for the development below. For a unified treatment of the discrete-time case and the continuous-time case we should make sure that, in continuous time, we choose $\mathcal{D}$ such that the above properties hold. In this book we simply take $\mathcal{D} = C^\infty(\mathbb{R}_+, \mathbb{R})$, the space of real-valued functions on $\mathbb{R}_+$ that can be differentiated arbitrarily often. It follows from the theory of differential equations (see, for instance, [73, Chapter 4]) that Properties 3.1– 3.3 hold for this choice. However, other choices for $\mathcal{D}$ are also possible; see the discussions in [68, Section XI.2] and [7, Section 5.1].

As a corollary of Property 3.3 we have the following lemma:

Lemma 3.2 Let $p(s) \in \mathbb{R}[s]$ and $m(s) \in \mathbb{R}[s]$. Then

$$\ker p(\sigma) \subset \ker m(\sigma) \tag{3.2}$$

if and only if there exists a polynomial $a(s) \in \mathbb{R}[s]$ such that

$$m(s) = a(s)p(s).$$

Proof The "if" part follows from the fact that multiplication of polynomials carries over to composition of operators. To prove the "only if" part let us assume that (3.2) holds. Let $g(s)$ be the greatest common divisor of $p(s)$ and $m(s)$. Then there exist $n_1(s) \in \mathbb{R}[s]$ and $n_2(s) \in \mathbb{R}[s]$ such that $p(s) = n_1(s)g(s)$ and $m(s) = n_2(s)g(s)$. It follows that

$$\ker g(\sigma) \subset \ker p(\sigma).$$

On the other hand, since $\mathbb{R}[s]$ has the Bézout property, there exist $x_1(s) \in \mathbb{R}[s]$ and $x_2(s) \in \mathbb{R}[s]$ such that $g(s) = x_1(s)p(s)+x_2(s)m(s)$. It now follows from the inclusion (3.2) that

$$\ker p(\sigma) \subset \ker g(\sigma)$$

so that we conclude that $\ker p(\sigma) = \ker g(\sigma)$. Because of Property 3.3, $\deg p(s) = \deg g(s)$ so that $n_1(s)$ has to be a nonzero constant $c \in \mathbb{R}$. The lemma is now proven by taking $a(s) = n_2(s)/c$. ◇

In the matrix case a differential operator $P(\sigma)$ taking $\mathcal{D}^q$ to $\mathcal{D}^g$ will be associated to a polynomial matrix $P(s) \in \mathbb{R}^{g \times q}[s]$. The above three properties translate into the following three lemmas:

Lemma 3.3 Let $P(s) \in \mathbb{R}^{g \times q}[s]$ be such that $P(\sigma)\boldsymbol{x} = 0$ for all $\boldsymbol{x} \in \mathcal{D}^q$. Then $P(s) = 0$.

Proof Follows immediately from Property 3.1.

Lemma 3.4 Let $P(s) \in \mathbb{R}^{g \times q}[s]$ have full row rank and let $\boldsymbol{x}_1 \in \mathcal{D}^g$. Then there exists $\boldsymbol{x}_2 \in \mathcal{D}^q$ such that

$$P(\sigma)\boldsymbol{x}_2 = \boldsymbol{x}_1. \tag{3.3}$$

Proof Let $P(s)$ be factorized as in Theorem 2.1:

$$P(s) = U(s)D(s)V(s)$$

where $D(s) = [\Delta(s) \quad 0]$ is the Smith form of $P(s)$ over $\mathbb{R}[s]$, $U(s)$ and $V(s)$ are $\mathbb{R}[s]$-unimodular matrices of sizes $g \times g$ and $q \times q$, respectively, and $\Delta(s) = \text{diag}\,(d_1(s), d_2(s), \ldots, d_r(s))$ with $d_i(s) \in \mathbb{R}[s]$. Note that $D(s)$ has the above form because of the fact that $P(s)$ has full row rank. Let $\boldsymbol{x}_1 \in \mathcal{D}^g$. Define $\tilde{\boldsymbol{x}}_1 := U^{-1}(\sigma)\boldsymbol{x}_1$. It follows from Property 3.2 that there exists $\tilde{\boldsymbol{x}}_2 \in \mathcal{D}^q$ such that $D(\sigma)\tilde{\boldsymbol{x}}_2 = \tilde{\boldsymbol{x}}_1$. Defining

$$\boldsymbol{x}_2 = V^{-1}(\sigma) \begin{bmatrix} \tilde{\boldsymbol{x}}_2 \\ 0 \end{bmatrix}$$

we have that $P(\sigma)\boldsymbol{x}_2 = U(\sigma)D(\sigma)\tilde{\boldsymbol{x}}_2 = U(\sigma)\tilde{\boldsymbol{x}}_1 = \boldsymbol{x}_1$.

Lemma 3.5 Let $P(s) \in \mathbb{R}^{q \times q}[s]$ be nonsingular. Then

$$\dim\ \ker P(\sigma) = \deg\det P(s). \tag{3.4}$$

Proof Let $P(s)$ be factorized as in Theorem 2.1:

$$P(s) = U(s)D(s)V(s)$$

where $D(s) = \text{diag}\,(d_1(s), d_2(s), \ldots, d_q(s))$ with $d_i(s) \in \mathbb{R}[s]$ is the Smith form of $P(s)$ over $\mathbb{R}[s]$ and both $U(s)$ and $V(s)$ are $\mathbb{R}[s]$-unimodular matrices of size $q \times q$. Then $\det P(s) = d_1(s)d_2(s)\cdots d_q(s)$ so that

$$\deg\det P(s) = \sum_{i=1}^{q} \deg d_i(s).$$

The mapping

$$\begin{aligned} \phi : \ \ker P(\sigma) &\rightarrow \ker D(\sigma) \\ \boldsymbol{x} &\mapsto V(\sigma)\boldsymbol{x} \end{aligned}$$

has an inverse ϕ^{-1} given by

$$\begin{aligned} \phi^{-1} : \ \ker D(\sigma) &\rightarrow \ker P(\sigma) \\ \boldsymbol{x} &\mapsto V^{-1}(\sigma)\boldsymbol{x}. \end{aligned}$$

As a result, ϕ is bijective so that we have

$$\dim \ \ker P(\sigma) = \dim \ \ker D(\sigma) = \sum_{i=1}^{q} \dim \ \ker d_i(\sigma) = \sum_{i=1}^{q} \deg d_i(s)$$

where the last equality follows from Property 3.3.

Corollary 3.6 Let $P(s) \in \mathbb{R}^{g \times q}[s]$ have full column rank for all $s \in \mathbb{C}$ and let $\boldsymbol{x} \in \mathcal{D}^q$. Then

$$P(\sigma)\boldsymbol{x} = 0 \implies \boldsymbol{x} = 0.$$

Proof From Theorem 2.1 it follows that we can find an $\mathbb{R}[s]$-unimodular matrix $U(s)$ of size $g \times g$ such that

$$U(s)P(s) = \begin{bmatrix} P_1(s) \\ 0 \end{bmatrix}$$

where $P_1(s) \in \mathbb{R}^{q \times q}[s]$ is a polynomial matrix of full rank. Then $P_1(s)$ has full rank for all $s \in \mathbb{C}$ so that det $P_1(s)$ is a nonzero constant. Because of the previous lemma we then have dim ker $P_1(\sigma) = 0$, from which the desired result follows. ◇

The following lemma is the matrix equivalent of Lemma 3.2.

Lemma 3.7 Let $P(s) \in \mathbb{R}^{g \times q}[s]$ and $M(s) \in \mathbb{R}^{k \times q}[s]$. Then

$$\ker P(\sigma) \subset \ker M(\sigma) \tag{3.5}$$

if and only if there exists a polynomial matrix $A(s) \in \mathbb{R}^{k \times g}[s]$ such that

$$M(s) = A(s)P(s).$$

Proof The "if" part follows as in the proof of Lemma 3.2. To prove the "only if" part let us assume that (3.5) holds. Denote the rank of $P(s)$ by r. Let $P(s)$ be factorized as in Theorem 2.1:

$$P(s) = U(s)D(s)V(s) \text{ with } D(s) = \begin{bmatrix} \Delta(s) & 0 \\ 0 & 0 \end{bmatrix}$$

where $D(s)$ is the Smith form of $P(s)$ over $\mathbb{R}[s]$ and both $U(s)$ and $V(s)$ are $\mathbb{R}[s]$-unimodular matrices. For an arbitrary row $m(s)$ of $M(s)$ we define the polynomial row vector $\tilde{m}(s) := m(s)V^{-1}(s)$. It is easily verified that (3.5) implies that

$$\ker D(\sigma) \subset \ker \tilde{m}(\sigma).$$

From this it follows that $\tilde{m}_i(s) = 0$ for $i = r+1, \ldots, q$ and that $\ker d_i(\sigma) \subset \ker \tilde{m}_i(\sigma)$ for $i = 1, \ldots, r$. According to Lemma 3.2 there exist $\tilde{a}_i(s) \in \mathbb{R}[s]$ such that $\tilde{m}_i(s) = \tilde{a}_i(s)d_i(s)$ for $i = 1, \ldots, r$. Defining

$$a(s) = [\tilde{a}_1(s) \cdots \tilde{a}_r(s) \quad 0 \quad 0]\, U^{-1}(s)$$

we have

$$\begin{aligned} m(s) &= \tilde{m}(s)V(s) \\ &= [\tilde{a}_1(s) \cdots \tilde{a}_r(s) \quad 0 \quad 0] \begin{bmatrix} \Delta(s) & 0 \\ 0 & 0 \end{bmatrix} V(s) \\ &= a(s)U(s) \begin{bmatrix} \Delta(s) & 0 \\ 0 & 0 \end{bmatrix} V(s) \\ &= a(s)P(s). \end{aligned}$$

By repeating the above for each row of $M(s)$, the desired result is obtained.

3.2 AR representations

In this book we restrict ourselves to the class of dynamical systems that is specified in the following definition.

Definition 3.8 The set $\mathcal{L}^q$ is defined as the set of dynamical systems $\Sigma = (\mathbf{T}, W, \mathcal{B})$ (where $W = \mathbb{R}^q$) for which the behavior $\mathcal{B}$ is the solution set of equations of the form:

$$R(\sigma)w = 0 \qquad (w \in \mathcal{D}^q \subset W^{\mathbf{T}}). \tag{3.6}$$

Here $R(s)$ is a polynomial matrix with coefficients in $\mathbb{R}$ and σ denotes either shift ($\mathbf{T} = \mathbb{Z}_+$) or differentiation ($\mathbf{T} = \mathbb{R}_+$); the set $\mathcal{D}^q$ denotes the set of functions to which elements of $\mathcal{B}$ should belong; see the remarks on page 40. ◇

Following the terminology of [66] we call the set of equations (3.6) an *AR representation* (*AutoRegressive representation*) of $\mathcal{B}$. It is easily verified that the behavior $\mathcal{B}$ of a system $\Sigma \in \mathcal{L}^q$ is a linear subspace of $W^{\mathbf{T}}$ and that the following holds for all $\tau \in \mathbf{T}$ (the system Σ is then called *time-invariant*):

$$w(.) \in \mathcal{B} \implies w(. + \tau) \in \mathcal{B}. \tag{3.7}$$

Note that in the case $\mathbf{T} = \mathbb{Z}_+$ the implication (3.7) is equivalent to $\sigma\mathcal{B} \subset \mathcal{B}$, i.e., σ-invariance of $\mathcal{B}$. For this case it has been shown in [66] that $\mathcal{B}$ is the behavior of a system in $\mathcal{L}^q$ if and only if $\mathcal{B}$ is a linear, σ-invariant and closed subspace of $W^{\mathbb{Z}_+}$. In continuous time the issue of characterizing subspaces of $\mathcal{D}^q$ that are kernels of polynomial differential operators is much more complicated; see the discussion in [68, XI-2]. It has recently been solved in [55].

Examples of AR representations are equations (1.1), (1.5) and the equations in Example 1.2.

The following question arises: what is the relation between polynomial matrices that represent the same behavior? The answer is essentially given by Lemma 3.7: the matrices are related via left multiplication by polynomial matrices. When the matrices that one starts with are of full row rank, the result can be formulated in a concise manner (see the next theorem). Note that we may restrict ourselves to matrices of full row rank because of the fact that for any polynomial matrix $R(s)$ we can find a matrix $\tilde{R}(s)$ of full row rank such that $\tilde{R}(s)$ represents the same behavior: from Theorem 2.1 it follows that we can find an $\mathbb{R}[s]$-unimodular matrix $U(s)$ such that

$$U(s)R(s) = \begin{bmatrix} \tilde{R}(s) \\ 0 \end{bmatrix}$$

where $\tilde{R}(s)$ has full row rank.

Theorem 3.9 (cf. [66, Section 4] and [53]) Let $\Sigma = (\mathbf{T}, W, \mathcal{B}) \in \mathcal{L}^q$ with $\mathcal{B}$ given by the AR representation

$$R(\sigma)w = 0.$$

Let $\tilde{\Sigma} = (\mathbf{T}, W, \tilde{\mathcal{B}}) \in \mathcal{L}^q$ with $\tilde{\mathcal{B}}$ given by the AR representation

$$\tilde{R}(\sigma)w = 0.$$

Assume that both $R(s) \in \mathbb{R}^{r\times q}[s]$ and $\tilde{R}(s) \in \mathbb{R}^{\tilde{r}\times q}[s]$ have full row rank. Then $\mathcal{B} = \tilde{\mathcal{B}}$ if and only if $r = \tilde{r}$ and there exists an $\mathbb{R}[s]$-unimodular matrix $U(s) \in \mathbb{R}^{r\times r}[s]$ such that $\tilde{R}(s) = U(s)R(s)$.

Proof The "if" part is immediate from Lemma 3.7. In order to prove the "only if" part let us assume that $\mathcal{B} = \tilde{\mathcal{B}}$. According to Lemma 3.7 there exist polynomial matrices $U(s)$ and $F(s)$ of sizes $\tilde{r} \times r$ and $r \times \tilde{r}$, respectively, with $\tilde{R}(s) = U(s)R(s)$ and $R(s) = F(s)\tilde{R}(s)$. As a result, $(I - F(s)U(s))R(s) = 0$, and it follows from the assumption that $R(s)$ has full row rank that $F(s)U(s) = I$. In an analogous way we also have that $U(s)F(s) = I$. We conclude that $r = \tilde{r}$ and that $U(s)$ is $\mathbb{R}[s]$-unimodular. $\diamond$

We now introduce a second parameter λ since we shall need to work with two parameters (λ and s) in Chapter 5. In the sequel we will use the notation that was introduced in Section 2.6: the space of rational W-valued functions in the parameter λ is denoted by $W(\lambda)$; the space of proper functions in $W(\lambda)$ is denoted by $W_\infty(\lambda)$; the mapping π_1 is defined as in (2.23), whereas π_- denotes the projection of $W(\lambda)$ onto $\lambda^{-1}W_\infty(\lambda)$, effected by "deleting the polynomial part".

Definition 3.10 Let $\Sigma = (\mathbf{T}, W, \mathcal{B}) \in \mathcal{L}^q$ and let

$$R(\sigma)\boldsymbol{w} = 0$$

be an AR representation of $\mathcal{B}$. The *Rational Behavioral space (RB space)* of Σ is defined as

$$\mathcal{W}_0 = \{w(\lambda) \in \lambda^{-1}W_\infty(\lambda) \mid R(\lambda)w(\lambda) \text{ is polynomial }\}.$$

We also define, for $k \geq 1$,

$$\mathcal{W}_k = \{w(\lambda) \in \lambda^{-1}W_\infty(\lambda) \mid \lambda^{-k}R(\lambda)w(\lambda) \text{ is polynomial }\}$$

and

$$\mathcal{W}_* = \{w(\lambda) \in \lambda^{-1}W_\infty(\lambda) \mid w(\lambda) \in \mathcal{W}_k \text{ for all } k \in \mathbb{Z}_+\}.$$

The *Driving Variable space (DV space)* of Σ is defined as

$$W^0 = \pi_1\mathcal{W}_*.$$

The *Rational Controllable space (RC space)* of Σ is defined as

$$\mathcal{C} = \{w(\lambda) \in W(\lambda) \mid R(\lambda)w(\lambda) = 0\} \quad (= \ker R(\lambda)). \tag{3.8}$$

◇

Because of Lemma 3.7 the spaces $\mathcal{W}_k$ ($k \in \mathbb{Z}_+$) are well-defined: polynomial matrices that represent the same behavior are related by left multiplication by polynomial matrices and therefore determine the same $\mathcal{W}_k$. It follows immediately from the definition that

$$\mathcal{W}_0 \supset \mathcal{W}_1 \supset \cdots$$

and that this sequence has $\mathcal{W}_*$ as its limit space; it is reached in finitely many steps. In the same way W^0 is the limit space of the sequence

$$W_0^0 \supset W_1^0 \supset \cdots$$

where $W_k^0 \subset W$ is defined, for $k \geq 0$, by

$$W_k^0 := \pi_1\mathcal{W}_k. \tag{3.9}$$

Obviously we have

$$\mathcal{W}_* = \{w(\lambda) \in \lambda^{-1} W_\infty(\lambda) \mid w(\lambda) \in \mathcal{C}\}$$

and

$$\mathcal{C} = \{w(\lambda) \in W(\lambda) \mid \exists N \in \mathbb{Z}_+ \text{ such that } \lambda^{-N} w(\lambda) \in \mathcal{W}_*\}.$$

Both $\mathcal{W}_*$ and $\mathcal{C}$ do not depend on the particular choice of $R(s)$. Note that

$$\dim_{\mathbb{R}(\lambda)} \mathcal{W}_* = \dim_{\mathbb{R}} W^0 = \dim_{\mathbb{R}(\lambda)} \mathcal{C} = q - \text{rank } R(s).$$

The last equality shows that polynomial matrices that represent the same behavior have the same rank.

Observability indices and controllability indices are classically introduced (cf. [71, 72]) at the level of standard state space representations

$$\begin{aligned} \sigma \boldsymbol{x} &= A\boldsymbol{x} + B\boldsymbol{u} \\ \boldsymbol{y} &= C\boldsymbol{x} + D\boldsymbol{u}. \end{aligned}$$

The observability indices are then defined as integer invariants related to the pair (A, C), whereas the controllability indices are defined as integer invariants related to the pair (A, B). In this book we take another approach by introducing these indices at the more general level of AR representations. In this way the indices more genuinely reflect properties of a dynamical system. It will be shown in Section 6.1 and Section 6.2 that the indices coincide with the classical notions when the system is represented in state space form.

Definition 3.11 Let $\Sigma = (\mathbf{T}, W, \mathcal{B}) \in \mathcal{L}^q$ and let

$$R(\sigma)\boldsymbol{w} = 0$$

be an AR representation of $\mathcal{B}$. The *observability indices* of Σ are defined as the left Wiener-Hopf indices of $R(s)$. The *order* of Σ, denoted by $\text{ord}(\Sigma)$, is defined as the sum of the observability indices of Σ. The *controllability indices* of Σ are defined as the minimal indices of the RC space $\mathcal{C}$ of Σ. The *controllability order* of Σ, denoted by $\text{C-ord}(\Sigma)$, is defined as the sum of the controllability indices of Σ. ◇

In the terminology of [66] our observability indices constitute a "shortest lag structure" for Σ. They do not depend on the particular choice of the matrix $R(s)$ since it follows from the proof of Theorem 2.19 (equation (2.14)) that the left Wiener-Hopf indices $\kappa_1, \ldots, \kappa_r$ of $R(s)$ satisfy

$$\sharp\{j \mid \kappa_j \geq k+1\} = \dim W_k^0 - \dim W^0 \tag{3.10}$$

(W_k^0 defined by (3.9)). Note that $\text{ord}(\Sigma)$ coincides with $\kappa^\ell(R)$, the left Wiener-Hopf order of $R(s)$ while $\text{C-ord}(\Sigma)$ equals the minimal order $\nu(\mathcal{C})$ of the RC space $\mathcal{C}$. It follows from Lemma 2.28 that

$$\text{C-ord}(\Sigma) \leq \text{ord}(\Sigma) \tag{3.11}$$

with equality if $R(s)$ has full row rank for all $s \in \mathbb{C}$. In that case the observability indices of Σ coincide with the minimal indices of the rational vector space $\text{im}\, R^T(s)$ (Lemma 2.28); we will then call the system "controllable" as in [67, 68]. It has the property that an arbitrary "past" trajectory of its behavior can be connected to any desired "future" trajectory of the behavior in finite time.

The next theorem is important since it shows that $\mathcal{W}_0$, the RB space of Σ, is a complete invariant for the equivalence relation on representations that is induced by the behavioral approach. In fact, our terminology "Rational Behavioral space" (Definition 3.10) is based upon this. The theorem provides a very useful tool for proving the equivalence of representations.

Theorem 3.12 Let $\Sigma = (\mathbf{T}, W, \mathcal{B})$ and $\tilde{\Sigma} := (\mathbf{T}, W, \tilde{\mathcal{B}})$ be dynamical systems in $\mathcal{L}^q$. Then

$$\mathcal{B} = \tilde{\mathcal{B}}$$

if and only if the RB spaces of Σ and $\tilde{\Sigma}$ are the same.

Proof The "only if" part is immediate from the fact that the RB space of a dynamical system is well-defined. To prove the "if" part, let us assume that the RB spaces of Σ and $\tilde{\Sigma}$ are the same. Then it follows that the RC spaces of Σ and $\tilde{\Sigma}$ are also the same. The reason for this is that the RC space $\mathcal{C}$ can be written in terms of the RB space $\mathcal{W}_0$ as follows:

$$\mathcal{C} := \{w(\lambda) \in W(\lambda) \mid \exists\, N \in \mathbb{Z}_+ \text{ such that } \lambda^{-k} w(\lambda) \in \mathcal{W}_0 \text{ for all } k \geq N\}.$$

Let us now assume that $R(s) \in \mathcal{R}^{r\times q}$ and $\tilde{R}(s) \in \mathcal{R}^{r\times q}$ are polynomial matrices of full row rank representing $\mathcal{B}$ and $\tilde{\mathcal{B}}$, respectively (they have the same rank since $\ker R(s) = \ker \tilde{R}(s)$). Let $R(s) = U(s)\,[\Delta(s) \quad 0]\,B(s)$ be factorized as in Theorem 2.19: $U(s)$ is an $\mathbb{R}[s]$-unimodular matrix of size $r \times r$, $B(s)$ is an $\mathbb{R}_\infty(s)$-unimodular matrix of size $q \times q$, and $[\Delta(s) \quad 0]$ is the left Wiener-Hopf form of $R(s)$. We then have

$$\Delta(s) = \text{diag}\,(s^{\kappa_1}, \ldots, s^{\kappa_r})$$

where $\kappa_1 \geq \kappa_2 \geq \cdots \geq \kappa_r$ are the left Wiener-Hopf indices of $R(s)$. Of course we have

$$R(s)B^{-1}(s)\begin{bmatrix} 0 \\ I_{(q-r)\times(q-r)} \end{bmatrix} = 0 \tag{3.12}$$

and

$$R(s)B^{-1}(s)\begin{bmatrix}\Delta^{-1}(s)\\0\end{bmatrix}=U(s). \tag{3.13}$$

Because of the fact that the RC spaces of Σ and $\tilde{\Sigma}$ are the same, so that ker $R(s) =$ ker $\tilde{R}(s)$, it follows from (3.12) that also

$$\tilde{R}(s)B^{-1}(s)\begin{bmatrix}0\\I_{(q-r)\times(q-r)}\end{bmatrix}=0. \tag{3.14}$$

Furthermore, it follows from (3.13) that the matrix

$$R(s)\pi_-(B^{-1}(s)\begin{bmatrix}\Delta^{-1}(s)\\0\end{bmatrix})$$

is polynomial. Since the RB spaces of Σ and $\tilde{\Sigma}$ coincide, we conclude that

$$\tilde{R}(s)B^{-1}(s)\begin{bmatrix}\Delta^{-1}(s)\\0\end{bmatrix} \tag{3.15}$$

is a polynomial matrix, say $M(s)$. Equations (3.14) and (3.15) together imply that

$$\tilde{R}(s)B^{-1}(s)\begin{bmatrix}\Delta^{-1}(s) & 0\\0 & I_{(q-r)\times(q-r)}\end{bmatrix}=[M(s)\quad 0]$$

so that

$$\tilde{R}(s)=M(s)\,[\Delta(s)\quad 0]\,B(s). \tag{3.16}$$

Next we define the polynomial matrix $F(s)$ by $F(s) := M(s)U^{-1}(s)$; we then conclude from (3.16) that

$$\begin{aligned}\tilde{R}(s) &= F(s)U(s)\,[\Delta(s)\quad 0]\,B(s)\\ &= F(s)R(s).\end{aligned}$$

As a consequence ker $R(\sigma) \subset$ ker $\tilde{R}(\sigma)$. In a similar way it can be proven that ker $\tilde{R}(\sigma) \subset$ ker $R(\sigma)$. We conclude that ker $R(\sigma) =$ ker $\tilde{R}(\sigma)$, i.e., $\mathcal{B} = \tilde{\mathcal{B}}$.

3.3 ARMA representations

In many situations it is convenient to work with representations that contain other variables besides the external variables. When representations are directly written down from first principles such as laws of physics, these variables (here called *internal variables*) arise naturally. The first-order representations that were presented in Chapter 1 are examples of representations with internal variables. In this section we consider representations with internal variables in their full generality: we deal with so-called ARMA (AutoRegressive Moving Average) representations as defined in [67, 68]. In section 3.4 we will consider the special case of first-order representations.

Definition 3.13 Let $\Sigma = (\mathbf{T}, W, \mathcal{B})$ be a dynamical system and let $Z = \mathbb{R}^n$. Let $P(s) \in \mathbb{R}^{g \times q}[s]$ and $Q(s) \in \mathbb{R}^{g \times n}[s]$. The set of equations

$$P(\sigma)w = Q(\sigma)\xi \qquad (w \in \mathcal{D}^q \subset W^{\mathbf{T}}, \xi \in \mathcal{D}^n \subset Z^{\mathbf{T}}) \tag{3.17}$$

is an *ARMA representation* of $\mathcal{B}$ if

$$\mathcal{B} = \pi_W \mathcal{B}^f.$$

Here $\mathcal{B}^f$ is the behavior of the system $\Sigma^f = (\mathbf{T}, W \times Z, \mathcal{B}^f) \in \mathcal{L}^{q+n}$ that is represented by (3.17), and the projection $\pi_W : W^{\mathbf{T}} \times Z^{\mathbf{T}} \to W^{\mathbf{T}}$ is given by $\pi_W(w, \xi) = w$; the set $\mathcal{D}^{q+n}$ denotes the set of functions to which elements of $\mathcal{B}^f$ should belong; see the remarks on $\mathcal{D}$ on page 40. $\diamond$

The next lemma follows immediately from Lemma 3.4.

Lemma 3.14 Let $\Sigma = (\mathbf{T}, W, \mathcal{B})$ with $\mathcal{B}$ given by the ARMA representation

$$\begin{bmatrix} P_1(\sigma) \\ P_2(\sigma) \end{bmatrix} w = \begin{bmatrix} Q_{11}(\sigma) & Q_{12}(\sigma) \\ Q_{21}(\sigma) & 0 \end{bmatrix} \begin{bmatrix} \xi_1 \\ \xi_2 \end{bmatrix}$$

where $Q_{12}(s)$ has full row rank. Then $\mathcal{B}$ is also represented by the ARMA representation

$$P_2(\sigma)w = Q_{21}(\sigma)\xi.$$

$\diamond$

Since an AR representation is a special case of an ARMA representation, the class $\mathcal{L}^q$ is contained in the class of dynamical systems whose behavior can be represented in ARMA form. The following lemma gives a procedure for rewriting ARMA representations in AR form and therefore shows that the two classes coincide.

Lemma 3.15 Let $\Sigma = (\mathbf{T}, W, \mathcal{B})$ with $\mathcal{B}$ given by the ARMA representation

$$P(\sigma)w = Q(\sigma)\xi.$$

Let $U_2(s)$ be a right $\mathbb{R}[s]$-unimodular matrix for which

$$\ker U_2(s) = \operatorname{im} Q(s).$$

Then an AR representation of $\mathcal{B}$ is given by

$$U_2(\sigma)P(\sigma)w = 0.$$

Proof Let $U_1(s)$ be a polynomial matrix such that

$$U(s) = \begin{bmatrix} U_1(s) \\ U_2(s) \end{bmatrix}$$

is $\mathbb{R}[s]$-unimodular. Then

$$\begin{bmatrix} U_1(s) \\ U_2(s) \end{bmatrix} Q(s) = \begin{bmatrix} T(s) \\ 0 \end{bmatrix}$$

where $T(s)$ is a polynomial matrix of full row rank. Clearly, an ARMA representation of $\mathcal{B}$ is given by

$$\begin{bmatrix} U_1(\sigma) \\ U_2(\sigma) \end{bmatrix} P(\sigma)\boldsymbol{w} = \begin{bmatrix} U_1(\sigma) \\ U_2(\sigma) \end{bmatrix} Q(\sigma)\boldsymbol{\xi}$$

which can also be written as

$$\begin{bmatrix} U_1(\sigma)P(\sigma) \\ U_2(\sigma)P(\sigma) \end{bmatrix} \boldsymbol{w} = \begin{bmatrix} T(\sigma) \\ 0 \end{bmatrix} \boldsymbol{\xi}.$$

The statement now follows immediately from the previous lemma. ◇

In the next lemma we express the rational spaces $\mathcal{W}_k$ $(k \in \mathbb{Z}_+)$, $\mathcal{W}_*$, and $\mathcal{C}$ that were introduced in the previous section in terms of the matrices of an ARMA representation.

Lemma 3.16 Let $\Sigma = (\mathbf{T}, W, \mathcal{B}) \in \mathcal{L}^q$ and let

$$P(\sigma)\boldsymbol{w} = Q(\sigma)\boldsymbol{\xi}$$

be an ARMA representation of $\mathcal{B}$. Then we have, for all $k \in \mathbb{Z}_+$,

(i) $\mathcal{W}_k = \{w(\lambda) \in \lambda^{-1}W_\infty(\lambda) \mid \exists \xi(\lambda) \in Z(\lambda)$ such that $\lambda^{-k}P(\lambda)w(\lambda) - Q(\lambda)\xi(\lambda)$ is polynomial$\}$

(ii) $\mathcal{W}_* = \{w(\lambda) \in \lambda^{-1}W_\infty(\lambda) \mid P(\lambda)w(\lambda) \in \text{ im } Q(\lambda)\}$

(iii) $\mathcal{C} = \{w(\lambda) \in W(\lambda) \mid P(\lambda)w(\lambda) \in \text{ im } Q(\lambda)\}$.

Proof Let $U_2(s)$ be as in Lemma 3.15 and let $R(s) = U_2(s)P(s)$. According to Lemma 3.15 the behavior $\mathcal{B}$ is given by the AR representation

$$R(\sigma)\boldsymbol{w} = 0.$$

To prove (i), let $w(\lambda) \in \lambda^{-1}W_\infty(\lambda)$ and $\xi(\lambda) \in Z(\lambda)$ be such that $\lambda^{-k}P(\lambda)w(\lambda) - Q(\lambda)\xi(\lambda)$ is polynomial. Then

$$\lambda^{-k}U_2(\lambda)P(\lambda)w(\lambda) - U_2(\lambda)Q(\lambda)\xi(\lambda) = \lambda^{-k}U_2(\lambda)P(\lambda)w(\lambda)$$

is also polynomial and this proves that $w(\lambda) \in \mathcal{W}_k$ (Definition 3.10). Conversely, let $w(\lambda) \in \mathcal{W}_k$ so that $w(\lambda) \in \lambda^{-1}W_\infty(\lambda)$ and

$\lambda^{-k}U_2(\lambda)P(\lambda)w(\lambda)$ is polynomial, say $p(\lambda)$. Let $L(\lambda)$ be the polynomial right inverse of $U_2(\lambda)$. Then $\lambda^{-k}U_2(\lambda)P(\lambda)w(\lambda) = U_2(\lambda)L(\lambda)p(\lambda)$. This implies $\lambda^{-k}P(\lambda)w(\lambda) - L(\lambda)p(\lambda) \in \ker U_2(\lambda)$ so there exists $\xi(\lambda) \in Z(\lambda)$ such that $\lambda^{-k}P(\lambda)w(\lambda) - L(\lambda)p(\lambda) = Q(\lambda)\xi(\lambda)$. Then $\lambda^{-k}P(\lambda)w(\lambda) - Q(\lambda)\xi(\lambda)$ is polynomial and we conclude that (i) holds. Statements (ii) and (iii) are proven in an analogous way. ◇

The behavioral approach induces a natural equivalence relation for ARMA representations:

Definition 3.17 The ARMA representations

$$P(\sigma)w = Q(\sigma)\xi$$

and

$$\tilde{P}(\sigma)w = \tilde{Q}(\sigma)\xi$$

are *strongly externally equivalent* if they represent the same behavior. ◇

We remark that strong external equivalence generalizes "input-output equivalence", as defined in [7] for systems in AR form (3.6) for which an a priori decomposition of the external variables in inputs and outputs is made.

The next lemma gives a result on the strong external equivalence of ARMA representations that proves very useful in the sequel.

Lemma 3.18 Consider the ARMA representation

$$P(\sigma)w = Q(\sigma)\xi \tag{3.18}$$

where $P(s) \in \mathbb{R}^{g\times q}[s]$ and $Q(s) \in \mathbb{R}^{g\times n}[s]$. Let $U(s) \in \mathbb{R}^{g\times g}[s]$ be $\mathbb{R}[s]$-unimodular and let $\tilde{Q}(s)$ be a polynomial matrix such that im $Q(s)$ = im $\tilde{Q}(s)$. Then the ARMA representation

$$U(\sigma)P(\sigma)w = U(\sigma)\tilde{Q}(\sigma)\xi \tag{3.19}$$

is strongly externally equivalent to (3.18).

Proof It follows from Lemma 3.16 that the RB spaces corresponding to (3.18) and (3.19) are the same. From Theorem 3.12 we then conclude that the representations are strongly externally equivalent. ◇

It is clear that the RC space $\mathcal{C}$ is far from being a complete invariant under strong external equivalence. The next definition therefore introduces a concept of equivalence that is substantially weaker than the previous one.

Definition 3.19 The ARMA representations

$$P(\sigma)w = Q(\sigma)\xi$$

and

$$\tilde{P}(\sigma)w = \tilde{Q}(\sigma)\xi$$

are *weakly externally equivalent* if the corresponding RC spaces are the same. ◇

It is important to notice that not only the space $\mathcal{C}$ is (trivially) invariant under weak external equivalence, but also the spaces $\mathcal{W}_*$ and W^0 as well as the controllability indices (Definition 3.11). To get an idea of the difference between strong external equivalence and weak external equivalence we consider the representations $w = 0$ and $(\sigma + 1)w = 0$. In a trivial way the representations are weakly externally equivalent; however, they are *not* strongly externally equivalent. In the present context of C^∞-behaviors ($\mathbb{R}^{\mathbb{Z}_+}$-behaviors) the concept of weak external equivalence is less natural[1]. However, we will see that results on duality of representations can be nicely formulated in terms of weak external equivalence. We remark that weak external equivalence coincides with equivalence notions that have been used before in the literature; both the equivalence relation used by Grimm [21] and the equivalence relation used by Aplevich [1]–[3] coincide with weak external equivalence: our space $\mathcal{C}$ equals the "external behavior" in [1]–[3].

An interesting class of representations is the class of ARMA representations (3.17) for which $P(s)$ is the identity matrix:

$$w = Q(\sigma)\xi. \tag{3.20}$$

Again following the terminology of Willems, we call such a representation a *Moving Average (MA) representation.* For MA representations the concepts of strong external equivalence and weak external equivalence coincide:

Theorem 3.20 MA representations are strongly externally equivalent if and only if they are weakly externally equivalent.

Proof The "only if" part follows immediately from the fact that $\mathcal{C}$ is invariant under strong external equivalence. In order to prove the "if" part, assume that the MA representations

$$w = Q(\sigma)\xi$$

[1] In the context of L_2-behaviors (ℓ_2-behaviors), where the set of functions $\mathcal{D}$ is chosen as the space of quadratically integrable (summable) functions, weak external equivalence arises naturally as the type of equivalence, according to which representations of the same behavior are related; see [63, Theorem 3.8].

and

$$w = \tilde{Q}(\sigma)\xi$$

are weakly externally equivalent. According to Lemma 3.16 we then have that

$$\text{im } Q(s) = \text{ im } \tilde{Q}(s)$$

and Lemma 3.18 yields the desired result. ◇

Because of the above theorem it will come as no surprise that the class of dynamical systems that can be given in MA form is strictly smaller than $\mathcal{L}^q$. We first present the following definition.

Definition 3.21 The set $\mathcal{L}^q_c$ is defined as the set of dynamical systems Σ in $\mathcal{L}^q$ for which the order and the controllability order coincide:

$$\text{ord}(\Sigma) = \text{C-ord}(\Sigma).$$

◇

The next theorem shows that $\mathcal{L}^q_c$ coincides with the class of systems that can be given in MA form. Consequently, $\mathcal{L}^q_c$ consists of systems that are controllable in the sense of Willems [67, 68]. In fact, our terminology "Rational Controllable space" (Definition 3.10) is based on this observation. The theorem also shows that for systems in $\mathcal{L}^q_c$ the observability indices and controllability indices play dual roles: the observability indices relate to the shortest "lags" in an AR representation of Σ, whereas the controllability indices relate to the shortest "lags" in an MA representation.

Theorem 3.22 Let $\Sigma = (\mathbf{T}, W, \mathcal{B})$ be a dynamical system. Then $\Sigma \in \mathcal{L}^q_c$ if and only if there exists a polynomial matrix $Q(s)$ such that $\mathcal{B}$ is given by the MA representation

$$w = Q(\sigma)\xi.$$

In that case the observability indices of Σ coincide with the minimal indices of ker $Q^T(s)$, whereas the controllability indices coincide with the minimal indices of im $Q(s)$. In particular, the controllability indices equal the column degrees of $Q(s)$ if $Q(s)$ is column reduced and has full column rank for all $s \in \mathbb{C}$.

Proof To prove the "if" part let us first assume that a polynomial matrix $Q(s)$ with the above properties exist. Without loss of generality (Lemma 3.18), we may assume that $Q(s)$ has full column rank. Let $U_2(s)$ be a right $\mathbb{R}[s]$-unimodular matrix for which

$$\ker U_2(s) = \text{ im } Q(s).$$

According to Lemma 3.15 an AR representation of $\mathcal{B}$ is given by

$$U_2(\sigma)w = 0.$$

The controllability indices of Σ are by definition equal to the minimal indices of ker $U_2(s)$ = im $Q(s)$. The last statement now follows from Lemma 2.25 and Lemma 2.28. Since $U_2(s)$ is right $\mathbb{R}[s]$-unimodular, it also follows from Lemma 2.28 that the observability indices of Σ coincide with the minimal indices of im $U_2^T(s)$ = ker $Q^T(s)$. This also yields

$$\mathrm{ord}(\Sigma) = \nu(\text{ ker } Q^T(s)) = c(Q^T)$$

where the last equality holds because of Lemma 2.28. On the other hand the same lemma implies that C-ord$(\Sigma) = \nu($ im $Q(s)) = c(Q)$. Since $c(Q) = c(Q^T)$ holds trivially, we conclude that ord(Σ) = C-ord(Σ). To prove the "only if" part let us assume that $\Sigma \in \mathcal{L}_c^q$, i.e., that ord$(\Sigma)$ = C-ord(Σ). Let $R(s)$ be an $r \times q$ polynomial matrix of full row rank such that

$$R(\sigma)w = 0$$

is an AR representation of $\mathcal{B}$. From the proof of Lemma 2.28 we have that

$$\text{C-ord}(\Sigma) = \mathrm{ord}(\Sigma) + \sum_{\beta \in \mathbb{C}} c_\beta(R)$$

which implies that $R(s)$ should have full row rank for all $s \in \mathbb{C}$. Consequently, there exists a polynomial matrix $\tilde{R}(s)$ such that

$$\begin{bmatrix} \tilde{R}(s) \\ R(s) \end{bmatrix}$$

is $\mathbb{R}[s]$-unimodular with polynomial inverse $\begin{bmatrix} Q(s) & \tilde{Q}(s) \end{bmatrix}$. By Lemma 3.14 $\mathcal{B}$ is also represented by the ARMA representation

$$\begin{bmatrix} \tilde{R}(s) \\ R(s) \end{bmatrix} w = \begin{bmatrix} 0 & I_{(q-r)\times(q-r)} \\ 0 & 0 \end{bmatrix} \begin{bmatrix} \xi_1 \\ \xi_2 \end{bmatrix}.$$

Multiplying both sides of this equation by $\begin{bmatrix} Q(s) & \tilde{Q}(s) \end{bmatrix}$, we conclude from Lemma 3.18 that $\mathcal{B}$ is represented by the MA representation

$$w = Q(\sigma)\xi$$

which proves the theorem. ◇

To each system $\Sigma = (\mathbf{T}, W, \mathcal{B})$ in $\mathcal{L}^q$ we can associate a system $\tilde{\Sigma} = (\mathbf{T}, W, \tilde{\mathcal{B}})$ in $\mathcal{L}_c^q$ as follows. Let $\mathcal{B}$ be given by the AR representation

$$R(\sigma)w = 0. \tag{3.21}$$

Then $\tilde{\mathcal{B}}$ is given by the MA representation

$$w = R^T(\sigma)\xi. \tag{3.22}$$

We have the following lemma.

Lemma 3.23 Let $\Sigma = (\mathbf{T}, W, \mathcal{B}) \in \mathcal{L}^q$ and let $\tilde{\Sigma} \in \mathcal{L}^q_c$ be defined as above. Then the controllability indices of Σ coincide with the observability indices of $\tilde{\Sigma}$. Moreover, the ARMA representation

$$P(\sigma)w = Q(\sigma)\xi \tag{3.23}$$

is weakly externally equivalent to the MA representation (3.22) (AR representation (3.21)) if and only if the ARMA representation

$$\begin{bmatrix} 0 \\ I \end{bmatrix} w = \begin{bmatrix} Q^T(\sigma) \\ P^T(\sigma) \end{bmatrix} \xi \tag{3.24}$$

is weakly externally equivalent to the AR representation (3.21) (MA representation (3.22)).

Proof The controllability indices of Σ are the minimal indices of the space ker $R(s)$. By Theorem 3.22 these coincide with the observability indices of $\tilde{\Sigma}$. The other statements essentially follow from Lemma 3.16: the RC spaces of (3.22) and (3.23) equal im $R^T(\lambda)$ and $\{w(\lambda) \in W(\lambda) \mid P(\lambda)w(\lambda) \in \text{im } Q(\lambda)\}$, respectively. The RC spaces of (3.21) and (3.24) equal ker $R(\lambda)$ and $P^T(\lambda)[\ \ker Q^T(\lambda)]$, respectively. The proof now follows from elementary results in linear algebra. The statements between brackets follow in an analogous way.

3.4 First-order representations

An ARMA representation that is first-order in the internal variables and zero-th order in the external variables is generally called a first-order representation. In this section we introduce two different kinds of first-order representations. The first of these is the so-called *pencil representation* (P *representation*):

$$\begin{bmatrix} 0 \\ I \end{bmatrix} w = \begin{bmatrix} \sigma G - F \\ H \end{bmatrix} \xi. \tag{3.25}$$

Here G, F and H are constant real-valued matrices that are interpreted as mappings; G and F are mappings from Z to X, and H is a mapping from Z to W. We shall denote the above representation by (F, G, H) and usually write (3.25) as

$$\sigma G\xi = F\xi \tag{3.26}$$

$$w = H\xi. \tag{3.27}$$

Our terminology stems from the fact that the dynamical equation (3.26) is determined by the matrix pencil $sG - F$.

Because of Lemma 3.15 we know how to obtain an AR representation that is strongly externally equivalent to a P representation (F, G, H). For the sake of clarity we shall repeat the result for this special case in the following lemma.

Lemma 3.24 Consider a P representation (F, G, H) where G and F have size $k \times n$ while H has size $q \times n$. Denote the rank of the matrix

$$\begin{bmatrix} sG - F \\ H \end{bmatrix} \tag{3.28}$$

by r. Let $V(s)$ and $R(s)$ be polynomial matrices of sizes $(k+q-r) \times k$ and $(k+q-r) \times q$, respectively, such that $[V(s) \quad R(s)]$ has full row rank for all $s \in \mathbb{C}$ and

$$\begin{bmatrix} V(s) & R(s) \end{bmatrix} \begin{bmatrix} sG - F \\ H \end{bmatrix} = 0.$$

Then an AR representation that is strongly externally equivalent to (F, G, H) is given by

$$R(\sigma)w = 0. \tag{3.29}$$

Moreover, the following implications hold:

(i) $sG - F$ has full row rank $\Longrightarrow$ $R(s)$ has full row rank

(ii) $sG - F$ has full row rank for all $s \in \mathbb{C}$ $\Longrightarrow$ $R(s)$ has full row rank for all $s \in \mathbb{C}$.

Proof The strong external equivalence of the representations follows immediately from Lemma 3.15. To prove (i) let us assume that $R(s)$ does not have full row rank. Then there exists an $\mathbb{R}[s]$-unimodular matrix $M(s)$ (use Theorem 2.1 for $\mathcal{R} = \mathbb{R}[s]$) such that

$$M(s)R(s) = \begin{bmatrix} \tilde{R}(s) \\ 0 \end{bmatrix}$$

where $\tilde{R}(s)$ has full row rank. Writing

$$M(s)V(s) = \begin{bmatrix} V_1(s) \\ V_2(s) \end{bmatrix}$$

we have

$$\begin{bmatrix} V_1(s) & \tilde{R}(s) \\ V_2(s) & 0 \end{bmatrix} \begin{bmatrix} sG - F \\ H \end{bmatrix} = \begin{bmatrix} 0 \\ 0 \end{bmatrix}.$$

The fact that $sG - F$ has full row rank implies $V_2(s) = 0$, which contradicts the fact that the matrix $M(s)\,[V(s) \quad R(s)]$ has full row rank. We conclude that $R(s)$ should have full row rank. Statement (ii) is proven in an analogous way.

Remark 3.25 The rows of the matrix $[V(s) \quad R(s)]$ form a polynomial basis for the left null space of (3.28). The basis is minimal in the sense of Definition 2.27 if $[V(s) \quad R(s)]$ is row reduced. $\diamond$

We now introduce a different type of first-order representation that we call the *dual pencil representation* (DP *representation*):

$$\sigma K\boldsymbol{\xi} = L\boldsymbol{\xi} + M\boldsymbol{w}. \tag{3.30}$$

Here K and L are mappings from X to Z, and M is a mapping from W to Z; we denote the representation by (K, L, M). Representations of this type have been considered before in [1, 67, 68].

As a corollary of Lemma 3.23, the following duality exists between P representations and DP representations; see also [3, Proposition 1.2]:

Lemma 3.26 Let $\Sigma = (\mathbf{T}, W, \mathcal{B}) \in \mathcal{L}^q$ and let $\tilde{\Sigma} \in \mathcal{L}^q_c$ be defined by the AR representation (3.21) and the MA representation (3.22), respectively. Then the DP representation (K, L, M) is weakly externally equivalent to the MA representation (3.22) (AR representation (3.21)) if and only if the P representation (F, G, H) with $F := L^T$, $G := K^T$, and $M := H^T$ is weakly externally equivalent to the AR representation (3.21) (MA representation (3.22)).

3.5 Systems with split external variables

As mentioned in Chapter 1 there are situations in which the external variables are split into two parts. We will use the notation $W = Y \oplus U$ where $Y = \mathbb{R}^p$ and $U = \mathbb{R}^m$. We do *not* adhere any special meaning to this decomposition of W; in particular we do not call trajectories $\boldsymbol{u} \in U^{\mathbf{T}}$ "input" trajectories, nor do we call trajectories $\boldsymbol{y} \in Y^{\mathbf{T}}$ "output" trajectories.

In Section 6.3 we investigate under what conditions a so-called input-output relation exists between $\boldsymbol{u}$ and $\boldsymbol{y}$. More precisely we have the following definition.

Definition 3.27 Let $\Sigma = (\mathbf{T}, Y \oplus U, \mathcal{B}) \in \mathcal{L}^{p+m}$ and let $\mathcal{C}$ be the RC space of Σ. The system Σ is defined to be an *input-output system* (*I-O system*) if $\mathcal{C}$ can be written as

$$\mathcal{C} = \operatorname{im} \begin{bmatrix} T(\lambda) \\ I_{m \times m} \end{bmatrix}$$

where $T(s) \in \mathbb{R}^{p \times m}(s)$ is called the *transfer function* of Σ. ◇

For I-O systems the transfer function is clearly a complete invariant under weak external equivalence. Consequently, for these systems weak external equivalence comes down to the well-known notion of transfer equivalence (according to which systems are equivalent if their transfer functions coincide). We remark that the controllability order of an I-O system Σ equals the McMillan degree of its transfer function $T(s)$ (Lemma 2.14):

$$\text{C-ord}(\Sigma) = \delta^p(T). \tag{3.31}$$

An example of an I-O system is a system whose behavior is given in standard state space form:

$$\begin{aligned} \sigma \boldsymbol{x} &= A\boldsymbol{x} + B\boldsymbol{u} \\ \boldsymbol{y} &= C\boldsymbol{x} + D\boldsymbol{u}. \end{aligned} \tag{3.32}$$

The transfer function is then given as $T(s) = C(sI - A)^{-1}B + D$.

Let us now return to systems that are not necessarily I-O systems. A generalized version of (3.32) is the so-called *descriptor representation* (D *representation*):

$$\begin{aligned} \sigma E\boldsymbol{\xi} &= A\boldsymbol{\xi} + B\boldsymbol{u} \\ \boldsymbol{y} &= C\boldsymbol{\xi} + D\boldsymbol{u}. \end{aligned} \tag{3.33}$$

Here E, A, B, C and D are constant matrices that are interpreted as mappings; E and A are mappings from X_d (descriptor space) to X_e (equation space), B is a mapping from U to X_d, C is a mapping from X_d to Y, and D is a mapping from U to Y. Note that we do not require E and A to be square. Also, there need not be a transfer function for the system. The representation (3.33) will be denoted by (E, A, B, C, D) and is written in ARMA form either as

$$\begin{bmatrix} 0 & 0 \\ I & 0 \\ 0 & I \end{bmatrix} \begin{bmatrix} \boldsymbol{y} \\ \boldsymbol{u} \end{bmatrix} = \begin{bmatrix} \sigma E - A & -B \\ C & D \\ 0 & I \end{bmatrix} \begin{bmatrix} \boldsymbol{\xi}_1 \\ \boldsymbol{\xi}_2 \end{bmatrix} \tag{3.34}$$

or as

$$\begin{bmatrix} 0 & B \\ I & -D \end{bmatrix} \begin{bmatrix} \boldsymbol{y} \\ \boldsymbol{u} \end{bmatrix} = \begin{bmatrix} \sigma E - A \\ C \end{bmatrix} \boldsymbol{\xi}. \tag{3.35}$$

Note that (3.34) is in pencil form while (3.35) is in dual pencil form. It will become clear in Chapter 5 that any system in $\mathcal{L}^{p+m}$ can be written in descriptor form.

A nice feature of the descriptor form is that it comprises both the pencil form and the dual pencil form (see [10]). Here the important thing to notice is that a P representation (F, G, H) corresponding to a system $\Sigma = (\mathbf{T}, W, \mathcal{B}) \in \mathcal{L}^q$ with $F, G : Z \to X$ can be interpreted as the D representation (E, A, B, C, D) corresponding to a system $\Sigma = (\mathbf{T}, W \oplus \{0\}, \mathcal{B}) \in \mathcal{L}^q$ where $E = G$, $A = F$, $B : \{0\} \to X$ is the zero mapping, $C = H$ and $D = 0$. In a similar way a DP representation (K, L, M) corresponding to a system $\Sigma = (\mathbf{T}, W, \mathcal{B}) \in \mathcal{L}^q$ with $K, L : X \to Z$ can be interpreted as the D representation (E, A, B, C, D) corresponding to a system $\Sigma = (\mathbf{T}, \{0\} \oplus W, \mathcal{B}) \in \mathcal{L}^q$ where $E = K$, $A = L$, $B = M$, $C : X \to \{0\}$ is the zero mapping and $D = 0$. This serves as a motivation to consider the class of D representations (E, A, B, C, D) with $D = 0$ separately. A representation in this class will be called a DZ representation (Descriptor with

Zero direct feedthrough matrix). It will be denoted by (E, A, B, C). The class of systems that can be given in DZ form is not smaller than the class of systems that can be given in D form: we will show in Subsection 3.5.3 that a D representation can always be rewritten as a DZ representation.

When the external variable space W is decomposed as $W = Y \oplus U$, a P representation is written as

$$\begin{aligned} \sigma G\xi &= F\xi \\ y &= H_y\xi \\ u &= H_u\xi \end{aligned}$$

and denoted by (F, G, H_y, H_u). In the following subsections we investigate the relation between the pencil form and the descriptor form. We will present algorithms for rewriting P representations in descriptor form (both D and DZ) and vice versa. It is our aim to use some of these algorithms in Section 4.2 to derive minimality results for descriptor representations; for this reason we will take care that no additional redundancies are created by the algorithms. The algorithms involve operations on matrices; if in a specific case some of these matrices are empty, the corresponding operations can, of course, be omitted. The results of the next subsections have been written down in [30, 33].

3.5.1 Relation $(F, G, H_y, H_u) \leftrightarrow (E, A, B, C, D)$

In this subsection we present two algorithms, one for obtaining a P representation (F, G, H_y, H_u) from a D representation (E, A, B, C, D) and the other for the reverse process. We already saw that a D representation (E, A, B, C, D) can be written in pencil form as (3.34). Since *all* the u variables are then introduced as internal variables, the representation (3.34) generally contains a number of redundant internal variables. An economic way of rewriting a D representation in pencil form is provided by the following algorithm. In the algorithm as few as possible u variables are introduced as internal variables.

Algorithm 3.28 Let a D representation be given by (E, A, B, C, D). Decompose the descriptor space X_d as $X_{d1} \oplus X_{d2}$ where $X_{d2} = \ker E$. Decompose the equation space X_e as $X_{e1} \oplus X_{e2} \oplus X_{e3}$ where $X_{e1} = \operatorname{im} E$ and $X_{e1} \oplus X_{e2} = \operatorname{im} [E \quad B]$. Accordingly write

$$E = \begin{bmatrix} E_{11} & 0 \\ 0 & 0 \\ 0 & 0 \end{bmatrix}, A = \begin{bmatrix} A_{11} & A_{12} \\ A_{21} & A_{22} \\ A_{31} & A_{32} \end{bmatrix}, B = \begin{bmatrix} B_1 \\ B_2 \\ 0 \end{bmatrix}, C = \begin{bmatrix} C_1 & C_2 \end{bmatrix}.$$

The matrix E_{11} is nonsingular, whereas the matrix B_2 has full row rank. By renumbering the u-variables if necessary, we can write

$$\begin{bmatrix} B_1 \\ B_2 \\ 0 \end{bmatrix} = \begin{bmatrix} B_{11} & B_{12} \\ B_{21} & B_{22} \\ 0 & 0 \end{bmatrix}, D = \begin{bmatrix} D_1 & D_2 \end{bmatrix}$$

where B_{22} is nonsingular. With respect to the above decompositions, the D representation (E, A, B, C, D) is given as

$$\begin{aligned} \sigma E_{11}\boldsymbol{\xi}_1 &= A_{11}\boldsymbol{\xi}_1 + A_{12}\boldsymbol{\xi}_2 + B_{11}\boldsymbol{u}_1 + B_{12}\boldsymbol{u}_2 \\ 0 &= A_{21}\boldsymbol{\xi}_1 + A_{22}\boldsymbol{\xi}_2 + B_{21}\boldsymbol{u}_1 + B_{22}\boldsymbol{u}_2 \\ 0 &= A_{31}\boldsymbol{\xi}_1 + A_{32}\boldsymbol{\xi}_2 \\ \boldsymbol{y} &= C_1\boldsymbol{\xi}_1 + C_2\boldsymbol{\xi}_2 + D_1\boldsymbol{u}_1 + D_2\boldsymbol{u}_2. \end{aligned} \tag{3.36}$$

Now define pencil matrices G, F, H_y and H_u by

$$\left[\begin{array}{c} sG - F \\ \hline H_y \\ \hline H_u \end{array}\right] := \left[\begin{array}{cccccc} I & -B_{12}B_{22}^{-1} & 0 & 0 & 0 \\ 0 & 0 & I & 0 & 0 \\ \hline 0 & D_2B_{22}^{-1} & 0 & I & 0 \\ \hline 0 & 0 & 0 & 0 & I \\ 0 & B_{22}^{-1} & 0 & 0 & 0 \end{array}\right] \times$$

$$\times \begin{bmatrix} sE_{11} - A_{11} & -A_{12} & -B_{11} \\ -A_{21} & -A_{22} & -B_{21} \\ -A_{31} & -A_{32} & 0 \\ C_1 & C_2 & D_1 \\ 0 & 0 & I \end{bmatrix}. \tag{3.37}$$

◇

In Section 4.2 we will show that the above algorithm preserves minimality properties. At this stage we only state the following lemma:

Lemma 3.29 Let (F, G, H_y, H_u) be a P representation that results from applying Algorithm 3.28 to a D representation (E, A, B, C, D). Then the two representations are strongly externally equivalent and we have

(i) rank G = rank E

(ii) dim ker G = dim ker E+ dim B^{-1}[im E]

(iii) $[E \quad B]$ has full row rank $\Longrightarrow$ G has full row rank

(iv) $\begin{bmatrix} E \\ C \end{bmatrix}$ has full column rank $\Longrightarrow$ $\begin{bmatrix} G \\ H \end{bmatrix}$ has full column rank

(v) $\begin{bmatrix} sE-A \\ C \end{bmatrix}$ has full column rank for all $s \in \mathbb{C} \implies \begin{bmatrix} sG-F \\ H \end{bmatrix}$ has full column rank for all $s \in \mathbb{C}$

(vi) $[sE-A \quad B]$ has full row rank for all $s \in \mathbb{C} \implies sG-F$ has full row rank for all $s \in \mathbb{C}$.

Proof The strong external equivalence is proven as follows. First, the behavior corresponding to (E, A, B, C, D) is not affected by a change of basis in X_d or X_e: a change of basis in X_e corresponds to left multiplication of both sides of (3.34) by a matrix of the form

$$\begin{bmatrix} M & 0 & 0 \\ 0 & I & 0 \\ 0 & 0 & I \end{bmatrix}$$

where M is nonsingular. A change of basis in X_d corresponds to right multiplication of the matrix on the right-hand side of (3.34) by a matrix of the form

$$\begin{bmatrix} N & 0 \\ 0 & I \end{bmatrix}$$

where N is nonsingular. These operations do not affect the behavior because of Lemma 3.18. As a result, (E, A, B, C, D) is strongly externally equivalent (up to a renumbering of u variables) to the representation (3.36), which can also be written as

$$\begin{bmatrix} 0 & 0 & 0 \\ 0 & 0 & 0 \\ 0 & 0 & 0 \\ I & 0 & 0 \\ 0 & I & 0 \\ 0 & 0 & I \end{bmatrix} \begin{bmatrix} \boldsymbol{y} \\ \boldsymbol{u}_1 \\ \boldsymbol{u}_2 \end{bmatrix} = \begin{bmatrix} \sigma E_{11} - A_{11} & -A_{12} & -B_{11} & -B_{12} \\ -A_{21} & -A_{22} & -B_{21} & -B_{22} \\ -A_{31} & -A_{32} & 0 & 0 \\ C_1 & C_2 & D_1 & D_2 \\ 0 & 0 & I & 0 \\ 0 & 0 & 0 & I \end{bmatrix} \begin{bmatrix} \xi_1 \\ \xi_2 \\ \xi_3 \\ \xi_4 \end{bmatrix}.$$

Left multiplication of both sides of the above equation by the nonsingular matrix

$$\begin{bmatrix} I & -B_{12}B_{22}^{-1} & 0 & 0 & 0 & 0 \\ 0 & 0 & I & 0 & 0 & 0 \\ 0 & B_{22}^{-1} & 0 & 0 & 0 & 0 \\ 0 & D_2 B_{22}^{-1} & 0 & I & 0 & 0 \\ 0 & 0 & 0 & 0 & I & 0 \\ 0 & B_{22}^{-1} & 0 & 0 & 0 & I \end{bmatrix}$$

leads to a strongly externally equivalent representation of the form

$$\begin{bmatrix} 0 & 0 \\ 0 & 0 \\ 0 & 0 \\ I & 0 \\ 0 & I \end{bmatrix} \begin{bmatrix} \boldsymbol{y} \\ \boldsymbol{u} \end{bmatrix} = \begin{bmatrix} \sigma G - F & 0 \\ * & -I \\ H_y & 0 \\ H_u & 0 \end{bmatrix} \begin{bmatrix} \xi_1 \\ \xi_2 \end{bmatrix}$$

where F, G, H_y and H_u are given by (3.37). Using Lemma 3.14 we conclude that this is strongly externally equivalent to the representation

$$\begin{bmatrix} 0 & 0 \\ I & 0 \\ 0 & I \end{bmatrix} \begin{bmatrix} \boldsymbol{y} \\ \boldsymbol{u} \end{bmatrix} = \begin{bmatrix} \sigma G - F \\ H_y \\ H_u \end{bmatrix} \boldsymbol{\xi}.$$

Next, (i) and (iii) are immediate. Denoting the number of columns of B_{21} by m_1, equality (ii) follows from

$$\begin{aligned} \dim \ker G &= \dim \ker E + m_1 \\ &= \dim \ker E + \dim \ker [B_{21} \quad B_{22}] \\ &= \dim \ker E + \dim B^{-1}[\operatorname{im} E]. \end{aligned}$$

Implication (iv) follows from

$$\dim \ker \begin{bmatrix} G \\ H \end{bmatrix} = \dim \ker \begin{bmatrix} I & 0 & 0 \\ 0 & I & -D_2 B_{22}^{-1} \\ 0 & 0 & -B_{22}^{-1} \end{bmatrix} \begin{bmatrix} E_{11} & 0 \\ C_1 & C_2 \\ A_{21} & A_{22} \end{bmatrix}.$$

The implications (v) and (vi) follow immediately from (3.37). ◇

We now return to the inverse problem of rewriting a P representation as a D representation. One method would be to write the P representation (F, G, H_y, H_u) as

$$\begin{aligned} \sigma \begin{bmatrix} G \\ 0 \end{bmatrix} \boldsymbol{\xi} &= \begin{bmatrix} F \\ H_u \end{bmatrix} \boldsymbol{\xi} + \begin{bmatrix} 0 \\ -I \end{bmatrix} \boldsymbol{u} \\ \boldsymbol{y} &= H_y \boldsymbol{\xi}. \end{aligned}$$

This representation generally contains a number of redundant internal variables since its equation space comprises the whole space U. An economic way of rewriting a P representation as a D representation is provided by the following algorithm.

Algorithm 3.30 Let a P representation be given by (F, G, H_y, H_u). Decompose the internal variable space Z as $Z_1 \oplus Z_2 \oplus Z_3$, where $Z_2 = \ker G \cap \ker H_u$, and $Z_2 \oplus Z_3 = \ker G$. Accordingly, write

$$\begin{aligned} G &= \begin{bmatrix} G_1 & 0 & 0 \end{bmatrix}, F = \begin{bmatrix} F_1 & F_2 & F_3 \end{bmatrix}, \\ H_y &= \begin{bmatrix} H_{y1} & H_{y2} & H_{y3} \end{bmatrix}, H_u = \begin{bmatrix} H_{u1} & 0 & H_{u3} \end{bmatrix}. \end{aligned}$$

The matrices G_1 and H_{u3} both have full column rank. By renumbering the $\boldsymbol{u}$-variables if necessary, we can write

$$H_{u1} = \begin{bmatrix} H_{11} \\ H_{21} \end{bmatrix}, H_{u3} = \begin{bmatrix} H_{13} \\ H_{23} \end{bmatrix}$$

where H_{23} is nonsingular. With respect to the above decompositions, the P representation (F, G, H_y, H_u) is given as

$$\begin{aligned} \sigma G_1 \boldsymbol{z}_1 &= F_1 \boldsymbol{z}_1 + F_2 \boldsymbol{z}_2 + F_3 \boldsymbol{z}_3 \\ \boldsymbol{y} &= H_{y1} \boldsymbol{z}_1 + H_{y2} \boldsymbol{z}_2 + H_{y3} \boldsymbol{z}_3 \\ \boldsymbol{u}_1 &= H_{11} \boldsymbol{z}_1 + H_{13} \boldsymbol{z}_3 \\ \boldsymbol{u}_2 &= H_{21} \boldsymbol{z}_1 + H_{23} \boldsymbol{z}_3. \end{aligned} \tag{3.38}$$

Now define a D representation (E, A, B, C, D) by

$$\left[\begin{array}{c|c} sE - A & -B \\ \hline C & D \end{array}\right] := \left[\begin{array}{cccc} sG_1 - F_1 & -F_2 & -F_3 & 0 \\ -H_{11} & 0 & -H_{13} & I \\ \hline H_{y1} & H_{y2} & H_{y3} & 0 \end{array}\right] \times$$

$$\times \left[\begin{array}{cc|cc} I & 0 & 0 & 0 \\ 0 & I & 0 & 0 \\ -H_{23}^{-1} H_{21} & 0 & 0 & H_{23}^{-1} \\ 0 & 0 & I & 0 \end{array}\right].$$

Lemma 3.31 Let (E, A, B, C, D) be a D representation that results from applying Algorithm 3.30 to a P representation (F, G, H_y, H_u). Then the two representations are strongly externally equivalent and we have the following equalities:

(i) $\operatorname{rank} E = \operatorname{rank} G$

(ii) $\dim \ker E = \dim (Y \cap H[\ker G]) + \dim \ker \begin{bmatrix} G \\ H \end{bmatrix}$

(iii) $\operatorname{codim} \operatorname{im} E = \operatorname{codim} (Y + H[\ker G]) + \operatorname{codim} \operatorname{im} G$.

Proof It follows as in the proof of Lemma 3.29 that (F, G, H_y, H_u) is strongly externally equivalent to the representation (3.38), which can also be written as

$$\begin{bmatrix} 0 & 0 & 0 \\ I & 0 & 0 \\ 0 & I & 0 \\ 0 & 0 & I \end{bmatrix} \begin{bmatrix} \boldsymbol{y} \\ \boldsymbol{u}_1 \\ \boldsymbol{u}_2 \end{bmatrix} = \begin{bmatrix} \sigma G_1 - F_1 & -F_2 & -F_3 \\ H_{y1} & H_{y2} & H_{y3} \\ H_{11} & 0 & H_{13} \\ H_{21} & 0 & H_{23} \end{bmatrix} \begin{bmatrix} \xi_1 \\ \xi_2 \\ \xi_3 \end{bmatrix}.$$

By Lemma 3.14 this is strongly externally equivalent to the representation

$$\begin{bmatrix} 0 & 0 & 0 \\ 0 & 0 & 0 \\ I & 0 & 0 \\ 0 & I & 0 \\ 0 & 0 & I \end{bmatrix} \begin{bmatrix} \boldsymbol{y} \\ \boldsymbol{u}_1 \\ \boldsymbol{u}_2 \end{bmatrix} = \begin{bmatrix} \sigma G_1 - F_1 & -F_2 & -F_3 & 0 \\ -H_{11} & 0 & -H_{13} & I \\ H_{y1} & H_{y2} & H_{y3} & 0 \\ H_{11} & 0 & H_{13} & 0 \\ H_{21} & 0 & H_{23} & 0 \end{bmatrix} \begin{bmatrix} \xi_1 \\ \xi_2 \\ \xi_3 \\ \xi_4 \end{bmatrix}.$$

Addition of the second equation to the fourth leads to the strongly externally equivalent representation

$$\begin{bmatrix} 0 & 0 & 0 \\ 0 & 0 & 0 \\ I & 0 & 0 \\ 0 & I & 0 \\ 0 & 0 & I \end{bmatrix} \begin{bmatrix} y \\ u_1 \\ u_2 \end{bmatrix} = \begin{bmatrix} \sigma G_1 - F_1 & -F_2 & -F_3 & 0 \\ -H_{11} & 0 & -H_{13} & I \\ H_{y1} & H_{y2} & H_{y3} & 0 \\ 0 & 0 & 0 & I \\ H_{21} & 0 & H_{23} & 0 \end{bmatrix} \begin{bmatrix} \xi_1 \\ \xi_2 \\ \xi_3 \\ \xi_4 \end{bmatrix}.$$

Right multiplication of the matrix on the right-hand side by the nonsingular matrix

$$\begin{bmatrix} I & 0 & 0 & 0 \\ 0 & I & 0 & 0 \\ -H_{23}^{-1}H_{21} & 0 & 0 & H_{23}^{-1} \\ 0 & 0 & I & 0 \end{bmatrix}$$

leads to the strongly externally equivalent representation

$$\begin{bmatrix} 0 & 0 \\ I & 0 \\ 0 & I \end{bmatrix} \begin{bmatrix} y \\ u \end{bmatrix} = \begin{bmatrix} \sigma E - A & -B \\ C & D \\ 0 & I \end{bmatrix} \begin{bmatrix} \xi_1 \\ \xi_2 \end{bmatrix}.$$

Next, equality (i) is trivial while (ii) follows from

$$\begin{aligned} \dim \ker E &= \dim Z_2 \\ &= \dim (\ker G \cap \ker H_u) \\ &= \dim \ker G - \dim H_u[\ker G] \\ &= \dim(\ker G \cap \ker H) + \dim H[\ker G] - \dim H_u[\ker G] \\ &= \dim (\ker G \cap \ker H) + \dim (Y \cap H[\ker G]). \end{aligned}$$

Denoting the number of rows of H_{13} by m_1 we have

$$\begin{aligned} m_1 &= m - \dim Z_3 \\ &= m - (\dim \ker G - \dim (\ker G \cap \ker H_u) \\ &= m - \dim H_u[\ker G] \\ &= \operatorname{codim} H_u[\ker G] \\ &= \operatorname{codim} (Y + H[\ker G]). \end{aligned}$$

Since $\operatorname{codim} \operatorname{im} E = \operatorname{codim} \operatorname{im} G + m_1$, this implies (iii).

3.5.2 Relation $(F, G, H_y, H_u) \leftrightarrow (E, A, B, C)$

In this subsection we present an algorithm for obtaining a DZ representation from a P representation. The inverse problem of rewriting a DZ representation (E, A, B, C) in pencil form is solved by applying Algorithm 3.28 to the D representation $(E, A, B, C, 0)$. It will be shown in Section 4.2 that minimality properties are then preserved.

We now present the algorithm for obtaining a DZ representation from a P representation; compared to Algorithm 3.30 the internal variable space has to be decomposed in four parts instead of three parts.

Algorithm 3.32 Let a P representation be given by (F, G, H_y, H_u). Decompose the internal variable space Z as $Z_1 \oplus Z_2 \oplus Z_3 \oplus Z_4$, where $Z_4 = \ker G \cap \ker H$, $Z_3 \oplus Z_4 = \ker G \cap \ker H_y$, and $Z_2 \oplus Z_3 \oplus Z_4 = \ker G$. Accordingly, write

$$\begin{aligned} G &= [G_1 \quad 0 \quad 0 \quad 0], F = [F_1 \quad F_2 \quad F_3 \quad F_4], \\ H_y &= [H_{y1} \quad H_{y2} \quad 0 \quad 0], H_u = [H_{u1} \quad H_{u2} \quad H_{u3} \quad 0]. \end{aligned}$$

The matrices G_1, H_{y2} and H_{u3} have full column rank. By renumbering the $\boldsymbol{u}$-variables if necessary, we can write

$$H_{u1} = \begin{bmatrix} H_{11} \\ H_{21} \end{bmatrix}, H_{u2} = \begin{bmatrix} H_{12} \\ H_{22} \end{bmatrix}, H_{u3} = \begin{bmatrix} H_{13} \\ H_{23} \end{bmatrix}$$

where H_{23} is nonsingular. With respect to the above decompositions the P representation (F, G, H_y, H_u) is given as

$$\begin{aligned} \sigma G_1 \boldsymbol{z}_1 &= F_1 \boldsymbol{z}_1 + F_2 \boldsymbol{z}_2 + F_3 \boldsymbol{z}_3 + F_4 \boldsymbol{z}_4 \\ \boldsymbol{y} &= H_{y1} \boldsymbol{z}_1 + H_{y2} \boldsymbol{z}_2 \\ \boldsymbol{u}_1 &= H_{11} \boldsymbol{z}_1 + H_{12} \boldsymbol{z}_2 + H_{13} \boldsymbol{z}_3 \\ \boldsymbol{u}_2 &= H_{21} \boldsymbol{z}_1 + H_{22} \boldsymbol{z}_2 + H_{23} \boldsymbol{z}_3. \end{aligned}$$

Now define a DZ representation (E, A, B, C) by

$$\left[\begin{array}{c|c} sE - A & -B \\ \hline C & 0 \end{array}\right] := \left[\begin{array}{ccccc} sG_1 - F_1 & -F_2 & -F_3 & -F_4 & 0 \\ -H_{11} & -H_{12} & -H_{13} & 0 & I \\ \hline H_{y1} & H_{y2} & 0 & 0 & 0 \end{array}\right] \times$$

$$\times \left[\begin{array}{ccc|cc} I & 0 & 0 & 0 & 0 \\ 0 & I & 0 & 0 & 0 \\ -H_{23}^{-1} H_{21} & -H_{23}^{-1} H_{22} & 0 & 0 & H_{23}^{-1} \\ 0 & 0 & I & 0 & 0 \\ 0 & 0 & 0 & I & 0 \end{array}\right].$$

Lemma 3.33 Let (E, A, B, C) be a DZ representation that results from applying Algorithm 3.32 to a P representation (F, G, H_y, H_u). Then the two representations are strongly externally equivalent and we have the following equalities:

(i) $\operatorname{rank} E = \operatorname{rank} G$

(ii) $\dim \ker E = \dim \pi_Y(H[\ker G]) + \dim \ker \begin{bmatrix} G \\ H \end{bmatrix}$

(iii) $\operatorname{codim} \operatorname{im} E = \operatorname{codim} (U \cap H[\ker G]) + \operatorname{codim} \operatorname{im} G$.

Proof The strong external equivalence of the representations follows as in Lemma 3.31. Next, equality (i) is trivial while (ii) follows from

$$\begin{aligned}
\dim \ker E &= \dim Z_2 + \dim Z_4 \\
&= \dim Z_2 \oplus Z_3 \oplus Z_4 - \dim Z_3 \oplus Z_4 + \dim Z_4 \\
&= \dim \ker G - \dim(\ker G \cap \ker H_y) + \dim(\ker G \cap \ker H) \\
&= \dim H_y[\ker G] + \dim(\ker G \cap \ker H).
\end{aligned}$$

Denoting the number of rows of H_{13} by m_1 we have

$$\begin{aligned}
m_1 &= m - \dim Z_3 \\
&= m - (\dim(\ker G \cap \ker H_y) - \dim(\ker G \cap \ker H)) \\
&= m - (\dim H[\ker G] - \dim H_y[\ker G]) \\
&= m - \dim(U \cap H[\ker G]) \\
&= \operatorname{codim}(U \cap H[\ker G]).
\end{aligned}$$

Statement (iii) now follows since codim im $E =$ codim im $G + m_1$.

3.5.3 Relation $(E, A, B, C, D) \leftrightarrow (E, A, B, C)$

In this subsection we first present an algorithm for obtaining a DZ representation from a D representation. The algorithm will not be used for proving minimality results, but simply serves the goal of making explicit how to eliminate a direct feedthrough matrix D in a descriptor representation.

Algorithm 3.34 Let a D representation be given by (E, A, B, C, D). Let $V = [V_1 \quad V_2]$ be an nonsingular matrix such that

$$D[V_1 \quad V_2] = [D_1 \quad 0]$$

where the matrix D_1 has full column rank.

Let $T = [T_1^T \quad T_2^T]^T$ be the inverse of V. Define a DZ representation $(\tilde{E}, \tilde{A}, \tilde{B}, \tilde{C})$ by

$$\tilde{E} := \begin{bmatrix} E & 0 \\ 0 & 0 \end{bmatrix}, \tilde{A} := \begin{bmatrix} A & 0 \\ 0 & I \end{bmatrix}, \tilde{B} := \begin{bmatrix} B \\ -T_1 \end{bmatrix}, \tilde{C} := [C \quad D_1].$$

Lemma 3.35 Let $(\tilde{E}, \tilde{A}, \tilde{B}, \tilde{C})$ be a DZ representation that results from applying Algorithm 3.34 to a D representation (E, A, B, C, D). Then the two representations are strongly externally equivalent.

Proof Note that $D = D_1 T_1$. By Lemma 3.14 (E, A, B, C, D) is strongly externally equivalent to the representation

$$\begin{bmatrix} 0 & 0 \\ 0 & 0 \\ I & 0 \\ 0 & I \end{bmatrix} \begin{bmatrix} y \\ u \end{bmatrix} = \begin{bmatrix} \sigma E - A & 0 & -B \\ 0 & -I & T_1 \\ C & 0 & D_1 T_1 \\ 0 & 0 & I \end{bmatrix} \begin{bmatrix} \xi_1 \\ \xi_2 \\ \xi_3 \end{bmatrix}. \tag{3.39}$$

Left multiplication of both sides of (3.39) by the nonsingular matrix

$$\begin{bmatrix} I & 0 & 0 & 0 \\ 0 & I & 0 & 0 \\ 0 & -D_1 & I & 0 \\ 0 & 0 & 0 & I \end{bmatrix}$$

leads, by Lemma 3.18, to the strongly externally equivalent representation

$$\begin{bmatrix} 0 & 0 \\ I & 0 \\ 0 & I \end{bmatrix} \begin{bmatrix} \boldsymbol{y} \\ \boldsymbol{u} \end{bmatrix} = \begin{bmatrix} \sigma\tilde{E} - \tilde{A} & -\tilde{B} \\ \tilde{C} & 0 \\ 0 & I \end{bmatrix} \begin{bmatrix} \boldsymbol{\xi}_1 \\ \boldsymbol{\xi}_2 \end{bmatrix}.$$

◇

We now return to the inverse problem of rewriting a DZ representation as a D representation. Of course a DZ representation (E, A, B, C) is trivially written as the D representation $(E, A, B, C, 0)$. However, a more economic way of rewriting (E, A, B, C) in D form is provided by the following algorithm:

Algorithm 3.36 Let a DZ representation be given by (E, A, B, C). Decompose the descriptor space X_d as $X_{d1} \oplus X_{d2} \oplus X_{d3}$ where $X_{d3} = A^{-1}[\operatorname{im} E] \cap \ker E$ and $X_{d2} \oplus X_{d3} = \ker E$. Decompose the equation space X_e as $X_{e1} \oplus X_{e2} \oplus X_{e3}$ where $X_{e1} = \operatorname{im} E$ and $X_{e2} = AX_{d2}$. Accordingly write

$$E = \begin{bmatrix} E_{11} & 0 & 0 \\ 0 & 0 & 0 \\ 0 & 0 & 0 \end{bmatrix}, \; A = \begin{bmatrix} A_{11} & A_{12} & A_{13} \\ A_{21} & A_{22} & 0 \\ A_{31} & 0 & 0 \end{bmatrix},$$

$$B = \begin{bmatrix} B_1 \\ B_2 \\ B_3 \end{bmatrix}, \; C = [C_1 \;\; C_2 \;\; C_3].$$

The matrices E_{11} and A_{22} are nonsingular. With respect to the above decompositions, (E, A, B, C) is given as

$$\begin{aligned} \sigma E_{11}\boldsymbol{\xi}_1 &= A_{11}\boldsymbol{\xi}_1 + A_{12}\boldsymbol{\xi}_2 + A_{13}\boldsymbol{\xi}_3 + B_1\boldsymbol{u} \\ 0 &= A_{21}\boldsymbol{\xi}_1 + A_{22}\boldsymbol{\xi}_2 + B_2\boldsymbol{u} \\ 0 &= A_{31}\boldsymbol{\xi}_1 + B_3\boldsymbol{u} \\ \boldsymbol{y} &= C_1\boldsymbol{\xi}_1 + C_2\boldsymbol{\xi}_2 + C_3\boldsymbol{\xi}_3. \end{aligned} \tag{3.40}$$

Now define a D representation $(\tilde{E}, \tilde{A}, \tilde{B}, \tilde{C}, \tilde{D})$ by

$$\left[\begin{array}{c|c} s\tilde{E} - \tilde{A} & -\tilde{B} \\ \hline \tilde{C} & \tilde{D} \end{array}\right] := \left[\begin{array}{cccc} I & -A_{12}A_{22}^{-1} & 0 & 0 \\ 0 & 0 & I & 0 \\ \hline 0 & C_2A_{22}^{-1} & 0 & I \end{array}\right] \left[\begin{array}{cc|c} sE_{11} - A_{11} & -A_{13} & -B_1 \\ -A_{21} & 0 & -B_2 \\ -A_{31} & 0 & -B_3 \\ C_1 & C_3 & 0 \end{array}\right].$$

Lemma 3.37 Let $(\tilde{E}, \tilde{A}, \tilde{B}, \tilde{C}, \tilde{D})$ be a D representation that results from applying Algorithm 3.36 to a DZ representation (E, A, B, C). Then the two representations are strongly externally equivalent and we have

$$\tilde{A}[\ker \tilde{E}] \subset \text{ im } \tilde{E}. \tag{3.41}$$

Moreover, the following implications hold:

(i) $[E \quad B]$ has full row rank $\Longrightarrow$ $\begin{bmatrix}\tilde{E} & \tilde{B}\end{bmatrix}$ has full row rank

(ii) $\begin{bmatrix} E \\ C \end{bmatrix}$ has full column rank $\Longrightarrow$ $\begin{bmatrix} \tilde{E} \\ \tilde{C} \end{bmatrix}$ has full column rank

(iii) $\begin{bmatrix} sE - A \\ C \end{bmatrix}$ has full column rank for all $s \in \mathbb{C}$ $\Longrightarrow$ $\begin{bmatrix} s\tilde{E} - \tilde{A} \\ \tilde{C} \end{bmatrix}$ has full column rank for all $s \in \mathbb{C}$

(iv) $[sE - A \quad B]$ has full row rank for all $s \in \mathbb{C}$ $\Longrightarrow$ $\begin{bmatrix} s\tilde{E} - \tilde{A} & \tilde{B} \end{bmatrix}$ has full row rank for all $s \in \mathbb{C}$.

Proof As in the proof of Lemma 3.29 we have that (E, A, B, C) is strongly externally equivalent to the representation (3.40), which can be written as

$$\begin{bmatrix} 0 & 0 \\ 0 & 0 \\ 0 & 0 \\ I & 0 \\ 0 & I \end{bmatrix} \begin{bmatrix} y \\ u \end{bmatrix} = \begin{bmatrix} \sigma E_{11} - A_{11} & -A_{12} & -A_{13} & -B_1 \\ -A_{21} & -A_{22} & 0 & -B_2 \\ -A_{31} & 0 & 0 & -B_3 \\ C_1 & C_2 & C_3 & 0 \\ 0 & 0 & 0 & I \end{bmatrix} \begin{bmatrix} \xi_1 \\ \xi_2 \\ \xi_3 \\ \xi_4 \end{bmatrix}.$$

By interchanging the second, third and fourth columns of the matrix on the right-hand side of the above equation, we get the strongly externally equivalent representation

$$\begin{bmatrix} 0 & 0 \\ 0 & 0 \\ 0 & 0 \\ I & 0 \\ 0 & I \end{bmatrix} \begin{bmatrix} y \\ u \end{bmatrix} = \begin{bmatrix} \sigma E_{11} - A_{11} & -A_{13} & -B_1 & -A_{12} \\ -A_{21} & 0 & -B_2 & -A_{22} \\ -A_{31} & 0 & -B_3 & 0 \\ C_1 & C_3 & 0 & C_2 \\ 0 & 0 & I & 0 \end{bmatrix} \begin{bmatrix} \xi_1 \\ \xi_2 \\ \xi_3 \\ \xi_4 \end{bmatrix}.$$

Left multiplication of both sides of this equation by the nonsingular matrix

$$\begin{bmatrix} I & -A_{12}A_{22}^{-1} & 0 & 0 & 0 \\ 0 & 0 & I & 0 & 0 \\ 0 & C_2 A_{22}^{-1} & 0 & I & 0 \\ 0 & I & 0 & 0 & 0 \\ 0 & 0 & 0 & 0 & I \end{bmatrix}$$

leads, by Lemma 3.18, to a strongly externally equivalent representation of the form

$$\begin{bmatrix} 0 & 0 \\ I & 0 \\ 0 & I \end{bmatrix} \begin{bmatrix} y \\ u \end{bmatrix} = \begin{bmatrix} \sigma\tilde{E} - \tilde{A} & -\tilde{B} & 0 \\ \tilde{C} & \tilde{D} & 0 \\ * & * & -A_{22} \\ 0 & I & 0 \end{bmatrix} \begin{bmatrix} \xi_1 \\ \xi_2 \\ \xi_3 \end{bmatrix}$$

where $\tilde{E}$, $\tilde{A}$, $\tilde{B}$, $\tilde{C}$ and $\tilde{D}$ are given as in Algorithm 3.36. Since the matrix A_{22} is nonsingular, it follows from Lemma 3.14 that (E, A, B, C) is strongly externally equivalent to $(\tilde{E}, \tilde{A}, \tilde{B}, \tilde{C}, \tilde{D})$. Finally, equation (3.41) and the implications (i–iv) are easily verified.

4

Minimality and transformation groups

In the previous chapter we saw examples of P representations that contain more internal variables than needed to represent the behavior of a system. As a matter of fact, we might distinguish between "small" and "large" P representations. In the first section of this chapter we consider the question: what do small P representations look like? Formulated differently, we ask ourselves: which parts of the pencil equations are essential for the representation of the behavior? We will specify what we call "minimal" P representations and characterize minimality in terms of the constant pencil matrices. The same issue will be investigated for D representations, DZ representations and DP representations in subsequent sections. We will give results under both strong and weak external equivalence. A new feature of the approach in these sections is that invariant lower bounds are derived for *all* items that are to be minimized.

A second question that will be adressed in Subsection 4.5.1 is: what is the relation between the constant matrices of minimal P representations that are strongly (weakly) externally equivalent? This question will be answered for D representations, DZ representations and DP representations in subsequent subsections. Some of the results of this chapter have been written down in [29, 30, 33].

4.1 Minimality of a P representation

In this section we investigate the minimality of a P representation

$$\begin{aligned} \sigma G\boldsymbol{\xi} &= F\boldsymbol{\xi} \\ \boldsymbol{w} &= H\boldsymbol{\xi}. \end{aligned} \tag{4.1}$$

As before, G and F are mappings from Z to X. The representation will be called *minimal* if both Z and X have least dimension in the class of equivalent P representations. A minimal P representation thus involves a minimum number of internal variables *and* a minimum number of equations.

In the following two subsections we investigate minimality under strong external equivalence and weak external equivalence.

4.1.1 Results under strong external equivalence

We start with a lemma which gives a necessary condition for minimality.

Lemma 4.1 If a P representation (F,G,H) is minimal under strong external equivalence, then G has full row rank.

Proof (see "step one" of the reduction algorithm in [53]) Let us assume that (F,G,H) is minimal under strong external equivalence. Let V^* be the subspace of Z that is the limit space of the iteration (2.24) applied to the pencil $sG-F$, i.e., the iteration

$$V^0 = Z, \quad V^{m+1} = F^{-1}GV^m.$$

Decompose Z as $Z = Z_1 \oplus Z_2$ where $Z_1 = V^*$, and decompose X as $X = X_1 \oplus X_2$ where $X_1 = GV^*$. With respect to these decompositions we may write (see Section 2.6)

$$sG - F = \begin{bmatrix} sG_{11} - F_{11} & sG_{12} - F_{12} \\ 0 & sG_{22} - F_{22} \end{bmatrix}$$

where $sG_{22} - F_{22}$ has full column rank for all $s \in \mathbb{C}$, and G_{11} is a matrix of full row rank. Writing H with respect to the decomposition of Z as $H = [H_1 \quad H_2]$ we have (use Lemma 3.18) that (F,G,H) is strongly externally equivalent to the representation

$$\begin{bmatrix} 0 \\ 0 \\ I \end{bmatrix} w = \begin{bmatrix} \sigma G_{11} - F_{11} & \sigma G_{12} - F_{12} \\ 0 & \sigma G_{22} - F_{22} \\ H_1 & H_2 \end{bmatrix} \begin{bmatrix} \xi_1 \\ \xi_2 \end{bmatrix}.$$

It now follows from Corollary 3.6 that the representation (F,G,H) is strongly externally equivalent to the P representation (F_{11}, G_{11}, H_1). Because of the assumed minimality of (F,G,H), the number of rows of the matrix $sG_{22} - F_{22}$ should then be zero, i.e., $GV^* = X$. This implies that G has full row rank. (Note that the reduction procedure produces a P representation for which G has full row rank.) ◇

The following lemma presents lower bounds for the items that are to be minimized. Recall that $\delta^z(M)$ is defined as the total zero multiplicity of a matrix M (Definition 2.7), whereas $\nu(\mathcal{X})$ denotes the minimal order of a rational vector space $\mathcal{X}$ (Definition 2.27).

Lemma 4.2 Let $\Sigma = (\mathbf{T}, W, \mathcal{B}) \in \mathcal{L}^q$ and let (F,G,H) be a P representation of $\mathcal{B}$; here $F,\ G : Z \to X$ and $H : Z \to W$. Let W^0 be the DV space of Σ. Then

(i) $\dim X \geq \text{ord}(\Sigma)$

(ii) $\dim Z \geq \text{ord}(\Sigma) + \dim W^0$.

Moreover, if G has full row rank, we have

(a) $\text{rank } G = \text{ord}(\Sigma) + \delta^z(\begin{bmatrix} sG - F \\ H \end{bmatrix}) + \nu(\ker \begin{bmatrix} sG - F \\ H \end{bmatrix})$

(b) $\dim \ker G = \dim W^0 + \dim \ker \begin{bmatrix} sG - F \\ H \end{bmatrix}$.

Proof We first prove statements (a) and (b). From Lemma 2.39 it follows that

$$\text{rank } G = \nu(\text{ im } \begin{bmatrix} sG - F \\ H \end{bmatrix}) + \delta^z(\begin{bmatrix} sG - F \\ H \end{bmatrix}) + \nu(\ker \begin{bmatrix} sG - F \\ H \end{bmatrix}),$$

so that statement (a) is proved if it is shown that

$$\nu(\text{ im } \begin{bmatrix} sG - F \\ H \end{bmatrix}) = \text{ord}(\Sigma). \quad (4.2)$$

Let $V(s)$ and $R(s)$ be polynomial matrices that are constructed from F, G and H as in the proof of Lemma 3.24. Then

$$\ker \begin{bmatrix} V(s) & R(s) \end{bmatrix} = \text{ im } \begin{bmatrix} sG - F \\ H \end{bmatrix}, \quad (4.3)$$

so that it suffices to prove that

$$\nu(\ker \begin{bmatrix} V(s) & R(s) \end{bmatrix}) = \text{ord}(\Sigma). \quad (4.4)$$

Because of the fact that $sG - F$ has full row rank, the matrix $R(s)$ also has full row rank (Lemma 3.24). Without loss of generality we may assume that $R(s)$ is row reduced, so that the row degrees of $R(s)$ sum up to $\kappa^\ell(R) = \text{ord}(\Sigma)$ (Lemma 2.22). Since G has full row rank, it follows from (4.3) that the highest degree in a row of $V(s)$ has to be smaller than the highest degree in the corresponding row of $R(s)$. This implies that the matrix $[V(s) \quad R(s)]$ is also row reduced and that the row degrees of $[V(s) \quad R(s)]$ also sum up to $\text{ord}(\Sigma)$. Because of the right $\mathbb{R}[s]$-unimodularity of $[V(s) \quad R(s)]$, we have that $\nu(\ker [V(s) \quad R(s)]) = \kappa^\ell([V(s) \quad R(s)])$ equals the sum of the row degrees of $[V(s) \quad R(s)]$ (Lemma 2.28 and Lemma 2.22). Consequently, (4.4) holds and this proves (a). Note that the rows of $[V(s) \quad R(s)]$ are a minimal polynomial basis (Definition 2.27) for the left null space of $\begin{bmatrix} sG^T - F^T & H^T \end{bmatrix}^T$.

Next, statement (b) follows from (use Lemma 3.16)

$$\begin{aligned} \dim \ker G &= \dim \ker (sG - F) \\ &= \dim H[\ker (sG-F)] + \dim \ker \begin{bmatrix} sG - F \\ H \end{bmatrix} \\ &= \dim \mathcal{C} + \dim \ \ker \begin{bmatrix} sG - F \\ H \end{bmatrix} \\ &= \dim W^0 + \dim \ \ker \begin{bmatrix} sG - F \\ H \end{bmatrix} \end{aligned}$$

where $\mathcal{C}$ is the RC space of Σ. In order to prove (i) and (ii) let (F, G, H) be a P representation with $G : Z \to X$ for which G does not necessarily have full row rank. Let $(\tilde{F}, \tilde{G}, \tilde{H})$ be the P representation that is obtained by applying the reduction procedure of Lemma 4.1 to (F, G, H). Then obviously $\dim X \geq \text{rank } \tilde{G}$ and $\dim Z \geq \text{rank } \tilde{G} + \dim \ \ker \tilde{G}$. From (a) and (b) we have that $\text{rank } \tilde{G} \geq \text{ord}(\Sigma)$ and $\dim \ker \tilde{G} \geq \dim W^0$, so that (i) and (ii) hold. ◇

We now present the main theorem of this subsection.

Theorem 4.3 Let $\Sigma = (\mathbf{T}, W, \mathcal{B}) \in \mathcal{L}^q$ and let (F, G, H) be a P representation of $\mathcal{B}$. Then (F, G, H) is minimal under strong external equivalence if and only if the following conditions hold:

(i) G has full row rank

(ii) $\begin{bmatrix} G \\ H \end{bmatrix}$ has full column rank

(iii) $\begin{bmatrix} sG - F \\ H \end{bmatrix}$ has full column rank for all $s \in \mathbb{C}$.

Moreover, in case of minimality, the lower bounds in Lemma 4.2 (i) and (ii) are reached.

Proof To prove the "only if" part let us assume that (F, G, H) is minimal under strong external equivalence. The necessity of (i) has already been proved in Lemma 4.1; we therefore assume that G has full row rank. We will now prove the necessity of (ii). Decompose Z as $Z = Z_1 \oplus Z_2$ where $Z_2 = \ker G \cap \ker H$, and decompose X as $X = X_1 \oplus X_2$ where $X_2 = FZ_2$. With respect to these decompositions we may write

$$\begin{bmatrix} sG - F \\ H \end{bmatrix} = \begin{bmatrix} sG_{11} - F_{11} & 0 \\ -F_{21} & -F_{22} \\ H_1 & 0 \end{bmatrix}$$

where the matrix F_{22} has full row rank. It follows from Lemma 3.18 that the representation

$$\begin{bmatrix} 0 \\ 0 \\ I \end{bmatrix} w = \begin{bmatrix} \sigma G_{11} - F_{11} & 0 \\ -F_{21} & -F_{22} \\ H_1 & 0 \end{bmatrix} \begin{bmatrix} \xi_1 \\ \xi_2 \end{bmatrix} \tag{4.5}$$

is strongly externally equivalent to (F, G, H). According to Lemma 3.14 the representation (4.5) is strongly externally equivalent to the P representation (F_{11}, G_{11}, H_1). Because of the assumed minimality of (F, G, H), the number of columns of the matrix F_{22} should then be zero. Consequently ker $G \cap$ ker $H = \{0\}$, i.e., (ii) holds. (Note that the reduction procedure produces a P representation for which (ii) holds.) Next, we prove the necessity of (iii). Let V^* be the subspace of Z that is the limit space of the iteration (2.24) applied to the pencil

$$\begin{bmatrix} sG - F \\ H \end{bmatrix}.$$

The iteration is then given as

$$V^0 = Z, \quad V^{m+1} = (F^{-1} G V^m) \cap \ker H.$$

Decompose Z as $Z = Z_1 \oplus Z_2$ where $Z_1 = V^*$, and decompose X as $X = X_1 \oplus X_2$ where $X_1 = GV^*$. With respect to these decompositions we may write (note that $FV^* \subset GV^*$ and that $V^* \subset$ ker H)

$$\left[\begin{array}{c} sG - F \\ \hline H \end{array}\right] = \left[\begin{array}{cc} sG_{11} - F_{11} & sG_{12} - F_{12} \\ 0 & sG_{22} - F_{22} \\ \hline 0 & H_2 \end{array}\right].$$

Then it follows from Theorem 2.31 that the matrix

$$\begin{bmatrix} sG_{22} - F_{22} \\ H_2 \end{bmatrix}$$

has full column rank for all $s \in \mathbb{C}$, whereas, by definition, G_{11} is a matrix of full row rank. By Lemma 3.18 the representation (F, G, H) is strongly externally equivalent to the representation

$$\begin{bmatrix} 0 \\ 0 \\ I \end{bmatrix} w = \begin{bmatrix} \sigma G_{11} - F_{11} & \sigma G_{12} - F_{12} \\ 0 & \sigma G_{22} - F_{22} \\ 0 & H_2 \end{bmatrix} \begin{bmatrix} \xi_1 \\ \xi_2 \end{bmatrix}.$$

It now follows from Lemma 3.14 that (F, G, H) is strongly externally equivalent to the P representation (F_{22}, G_{22}, H_2). Because of the assumed minimality of (F, G, H), the number of columns of the matrix $sG_{11} - F_{11}$ should then be zero, i.e., $V^* = \{0\}$. By Theorem 2.31 this implies that (iii) holds.

(Note that the reduction procedure produces a P representation for which (iii) holds.)

Conversely, for a P representation (F, G, H) that satisfies conditions (i) and (ii), it follows from Lemma 2.36 that

$$\begin{bmatrix} sG - F \\ H \end{bmatrix} \text{ has full column rank and } \delta^z_\infty(\begin{bmatrix} sG - F \\ H \end{bmatrix}) = 0.$$

Also, condition (iii) implies that

$$\delta^z_\beta(\begin{bmatrix} sG - F \\ H \end{bmatrix}) = 0 \text{ for all } \beta \in \mathbb{C}.$$

Consequently, the lower bounds in Lemma 4.2 are reached. Since these lower bounds are invariant under strong external equivalence, the representation is minimal under strong external equivalence. ◇

In [65] minimality conditions under strong external equivalence are derived by Willems for P representations (F, G, H) with $G = [I \quad 0]$ (these are called "driving variable representations" in [68]). It is not difficult to check that the minimality conditions in [65] coincide with conditions (ii) and (iii) in the above theorem; see the remarks in [29]. In fact, our terminology "Driving Variable space" (Definition 3.10) is based on this observation.

Example 4.4 Consider the two interconnected mass-spring-damper systems of Example 1.2. Assume that $M_1 = M_2 = 1$, $c_1 = c_2 = 1$, and $k_1 = k_2 = 1$. A P representation for this system is easily obtained (see Example 1.3) as

$$\begin{bmatrix} 1 & 0 & 0 & 0 & 0 & 0 \\ 0 & 1 & 0 & 0 & 0 & 0 \\ 0 & 0 & 0 & 1 & 0 & 0 \\ 0 & 0 & 0 & 0 & 1 & 0 \\ 0 & 0 & 0 & 0 & 0 & 0 \\ 0 & 0 & 0 & 0 & 0 & 0 \end{bmatrix} \begin{bmatrix} \dot{\xi}_1 \\ \dot{\xi}_2 \\ \dot{\xi}_3 \\ \dot{\xi}_4 \\ \dot{\xi}_5 \\ \dot{\xi}_6 \end{bmatrix} = \begin{bmatrix} -1 & 1 & 0 & 0 & 0 & 0 \\ -1 & 0 & 1 & 0 & 0 & 0 \\ 0 & 0 & 0 & -1 & 1 & 0 \\ 0 & 0 & 0 & -1 & 0 & 1 \\ 1 & 0 & 0 & -1 & 0 & 0 \\ 0 & 0 & 1 & 0 & 0 & 1 \end{bmatrix} \begin{bmatrix} \xi_1 \\ \xi_2 \\ \xi_3 \\ \xi_4 \\ \xi_5 \\ \xi_6 \end{bmatrix}$$

$$\begin{bmatrix} \boldsymbol{q}_1 \\ \boldsymbol{F}_1 \\ \boldsymbol{q}_2 \\ \boldsymbol{F}_2 \end{bmatrix} = \begin{bmatrix} 1 & 0 & 0 & 0 & 0 & 0 \\ 0 & 0 & 1 & 0 & 0 & 0 \\ 0 & 0 & 0 & 1 & 0 & 0 \\ 0 & 0 & 0 & 0 & 0 & 1 \end{bmatrix} \begin{bmatrix} \xi_1 \\ \xi_2 \\ \xi_3 \\ \xi_4 \\ \xi_5 \\ \xi_6 \end{bmatrix}.$$

Since condition (i) of Theorem 4.3 fails, this is not a minimal P representation. Indeed, we will see in Chapter 5 that all minimal P representations of this system have only two ξ-variables; an example is

$$\begin{bmatrix} 1 & 0 \\ 0 & 1 \end{bmatrix} \begin{bmatrix} \dot{\xi}_1 \\ \dot{\xi}_2 \end{bmatrix} = \begin{bmatrix} -1 & 1 \\ -1 & 0 \end{bmatrix} \begin{bmatrix} \xi_1 \\ \xi_2 \end{bmatrix}$$

$$\begin{bmatrix} \boldsymbol{q}_1 \\ \boldsymbol{F}_1 \\ \boldsymbol{q}_2 \\ \boldsymbol{F}_2 \end{bmatrix} = \begin{bmatrix} 1 & 0 \\ 0 & 0 \\ 1 & 0 \\ 0 & 0 \end{bmatrix} \begin{bmatrix} \xi_1 \\ \xi_2 \end{bmatrix}.$$

◇

We conclude this subsection with a result that will be needed in Section 4.2. The following lemma expresses the DV space W^0 in terms of the matrices F, G and H of a P representation (F, G, H).

Lemma 4.5 Let $\Sigma = (\mathbf{T}, W, \mathcal{B}) \in \mathcal{L}^q$ and let $\mathcal{B}$ be given by a P representation (F, G, H) for which conditions (i) and (ii) of Theorem 4.3 hold. Then the DV space W^0 is given by

$$W^0 = H[\ker G].$$

Proof First we prove that $W^0 \subset H[\ker G]$. Let $w \in W^0$ and let $w(\lambda) \in \mathcal{W}_*$ be such that $w = \pi_1(w(\lambda))$. It follows from Lemma 3.16 that there exists $\xi(\lambda) \in Z(\lambda)$ such that

$$\begin{bmatrix} 0 \\ I \end{bmatrix} w(\lambda) = \begin{bmatrix} \lambda G - F \\ H \end{bmatrix} \xi(\lambda).$$

Since we assumed that $\ker G \cap \ker H = \{0\}$, it follows that $\xi(\lambda)$ is strictly proper and that its leading coefficient ξ_1 satisfies $G\xi_1 = 0$. Moreover, we have $w = H\xi_1$, and this implies that $W^0 \subset H[\ker G]$. Observe that it follows from Lemma 2.36 that the matrix

$$\begin{bmatrix} sG - F \\ H \end{bmatrix}$$

has full column rank. Using Lemma 3.16 and the assumption that G has full row rank we conclude that the desired result holds by an additional dimensionality argument:

$$\begin{aligned} \dim W^0 &= \dim \mathcal{C} \\ &= \dim H[\ker (sG - F)] \\ &= \dim \ker (sG - F) \\ &= \dim \ker G \\ &= \dim H[\ker G]. \end{aligned}$$

Remark 4.6 It follows from results in Section 6.3 that we actually have the following situation:

(i) $\operatorname{im} F \subset \operatorname{im} G \Longrightarrow H[\ker G] \subset W^0$

(ii) $F^{-1}[\operatorname{im} G] \cap \ker G \cap \ker H = \{0\} \Longrightarrow H[\ker G] \supset W^0$.

4.1.2 Results under weak external equivalence

In this subsection we consider the minimality issue of the previous subsection under weak external equivalence. The minimality conditions of Theorem 4.3 are, of course, necessary for minimality under weak external equivalence. However, they are not sufficient, which becomes clear from the following theorem.

Theorem 4.7 Let $\Sigma = (\mathbf{T}, W, \mathcal{B}) \in \mathcal{L}^q$ and let (F, G, H) be a P representation of $\mathcal{B}$; here $F, G : Z \to X$ and $H : Z \to W$. Let W^0 be the DV space of Σ. Then (F, G, H) is minimal under weak external equivalence if and only if the following conditions hold:

(i) G has full row rank

(ii) $\begin{bmatrix} G \\ H \end{bmatrix}$ has full column rank

(iii) $\begin{bmatrix} sG - F \\ H \end{bmatrix}$ has full column rank for all $s \in \mathbb{C}$

(iv) $sG - F$ has full row rank for all $s \in \mathbb{C}$.

Moreover, in case of minimality we have

(a) $\dim Z = \text{C-ord}(\Sigma) + \dim W^0$

(b) $\dim X = \text{C-ord}(\Sigma)$.

Proof To prove the "only if" part, let us assume that the P representation (F, G, H) is minimal under weak external equivalence. Then (F, G, H) should satisfy conditions (i)-(iii) because of Theorem 4.3 and the fact that weak external equivalence is weaker than strong external equivalence. We will now prove the necessity of condition (iv). Let T^* be the subspace of Z that is the limit space of the iteration (2.29) applied to the pencil $sG - F$, i.e.,

$$T^0 = \{0\}, \quad T^{m+1} = G^{-1}FT^m.$$

Decompose Z as $Z = Z_1 \oplus Z_2$ where $Z_1 = T^*$, and decompose X as $X = X_1 \oplus X_2$ where $X_1 = FT^*$. With respect to these decompositions we may write (note that $GT^* \subset FT^*$)

$$\left[\begin{array}{c} sG - F \\ \hline H \end{array}\right] = \left[\begin{array}{cc} sG_{11} - F_{11} & sG_{12} - F_{12} \\ 0 & sG_{22} - F_{22} \\ \hline H_1 & H_2 \end{array}\right].$$

It follows from Theorem 2.33 that the matrix $sG_{11} - F_{11}$ has full row rank for all $s \in \mathbb{C}$. Since $\ker G \subset T^*$, the matrix G_{22} has full column rank. Using Lemma 3.16 we now have that the RC space $\mathcal{C}$ of Σ is given as

$$\mathcal{C} = H[\ker(sG - F)] = H_1[\ker(sG_{11} - F_{11})].$$

As a consequence, the P representation (F, G, H) is weakly externally equivalent to the P representation (F_{11}, G_{11}, H_1). Because of the assumed minimality of (F, G, H), the number of rows of the matrix $sG_{22} - F_{22}$ should then be zero, i.e., $FT^* = X$. By Theorem 2.33 this implies (iv). (Note that the reduction procedure produces a P representation which satisfies (iv).)

Conversely, let (F, G, H) be a P representation that satisfies conditions (i)-(iv). Let $V(s)$ and $R(s)$ be polynomial matrices as in the proof of Theorem 4.2. It follows from Lemma 3.24 that $R(s)$ has full row rank for all $s \in \mathbb{C}$, so that by Lemma 2.28

$$\kappa^{\ell}(R) = \nu(\ker R(s)) = \text{C-ord}(\Sigma).$$

As a result (see the proof of Theorem 4.2), we have that rank $G = \kappa^{\ell}(R) =$ C-ord(Σ), and this proves (a). Furthermore, as in the proof of Theorem 4.2, we have that

$$\dim\ \ker G = \dim H[\ker(sG - F)] = \dim W^0$$

and that in general $\dim X \geq$ C-ord(Σ) and $\dim Z \geq$ C-ord$(\Sigma) + \dim W^0$. From the fact that C-ord(Σ) and W^0 are invariant under weak external equivalence we now conclude that (F, G, H) is minimal under weak external equivalence and that (a) and (b) hold.

4.2 Minimality of a D representation

In this section we investigate the minimality of a D representation

$$\begin{aligned} \sigma E\xi &= A\xi + Bu \\ y &= C\xi + Du. \end{aligned}$$

As before, E and A are mappings from X_d to X_e. The representation will be called *minimal* if both X_d and X_e have least dimension in the class of equivalent D representations. A minimal D representation thus involves a minimum number of internal variables *and* a minimum number of equations. The results in this section are based on the results for P representations in the previous section; the link is provided by Algorithm 3.28.

In order to put things into perspective, let us consider the restricted class of D representations (E, A, B, C, D) for which E is invertible. We are then essentially dealing with I-O systems (see Section 3.5) that have a proper transfer function. As remarked at the beginning of Section 3.5, for such systems the notion of weak external equivalence comes down to "transfer equivalence". From classical results for standard state space representations (see, for instance, [26, Theorem 6.2-3]), we conclude that the following conditions are necessary and sufficient for minimality under weak external equivalence in the restricted class:

(i) $\begin{bmatrix} sE - A \\ C \end{bmatrix}$ has full column rank for all $s \in \mathbb{C}$

(observability at all finite modes)

(ii) $[sE - A \quad B]$ has full row rank for all $s \in \mathbb{C}$

(controllability at all finite modes).

Under strong external equivalence the situation is different. It follows from a result in [65] that within the restricted class of D representations (E, A, B, C, D), for which E is invertible, the above condition (i) is not only necessary, but also sufficient. Hence controllability is not required for minimality under strong external equivalence. This is due to the fact that $\boldsymbol{y}$-trajectories that are associated to zero $\boldsymbol{u}$-trajectories are invariant under strong external equivalence (they are included in the behavior of the system), whereas they can be removed under weak external equivalence.

Let us now return to the general setting and investigate minimality for D representations (E, A, B, C, D) for which E can not only be singular, but also nonsquare. In the following two subsections we consider the situation under strong external equivalence and weak external equivalence, respectively.

4.2.1 Results under strong external equivalence

In analogy with the standard state space case, one would expect that observability at all finite modes (condition (i) above) is required for minimality under strong external equivalence. Indeed, we will see that (i) turns out to be a necessary condition. However, the following lemma shows that more conditions need to be satisfied for minimality.

Lemma 4.8 If a D representation (E, A, B, C, D) is minimal under strong external equivalence, then the following conditions hold:

(i) $[E \quad B]$ has full row rank

(ii) $\begin{bmatrix} E \\ C \end{bmatrix}$ has full column rank

(iii) $A[\ker E] \subset \operatorname{im} E$.

Proof Assume that (E, A, B, C, D) is minimal under strong external equivalence. Write (E, A, B, C, D) in ARMA form as

$$\begin{bmatrix} 0 & 0 \\ I & 0 \\ 0 & I \end{bmatrix} \begin{bmatrix} y \\ u \end{bmatrix} = \begin{bmatrix} \sigma E - A & -B \\ C & D \\ 0 & I \end{bmatrix} \begin{bmatrix} \xi_1 \\ \xi_2 \end{bmatrix}. \tag{4.6}$$

To prove (i), let us decompose X_d as $X_d = X_{d_1} \oplus X_{d_2}$ and X_e as $X_e = X_{e_1} \oplus X_{e_2}$ with $X_{d_2} = A^{-1}(\text{ im } [E \quad B])$ and $X_{e_1} = \text{ im } [E \quad B]$. With respect to these decompositions, the matrix

$$\begin{bmatrix} sE - A & -B \\ C & D \end{bmatrix}$$

has the form

$$\begin{bmatrix} sE_{11} - A_{11} & sE_{12} - A_{12} & -B_1 \\ -A_{21} & 0 & 0 \\ C_1 & C_2 & D \end{bmatrix}$$

where A_{21} has full column rank. Using Lemma 3.18 we have that (4.6) is strongly externally equivalent to the representation

$$\begin{bmatrix} 0 & 0 \\ 0 & 0 \\ I & 0 \\ 0 & I \end{bmatrix} \begin{bmatrix} y \\ u \end{bmatrix} = \begin{bmatrix} \sigma E_{11} - A_{11} & \sigma E_{12} - A_{12} & -B_1 \\ -A_{21} & 0 & 0 \\ C_1 & C_2 & D \\ 0 & 0 & I \end{bmatrix} \begin{bmatrix} \xi_1 \\ \xi_2 \\ \xi_3 \end{bmatrix},$$

which is by Corollary 3.6 strongly externally equivalent to $(E_{12}, A_{12}, B_1, C_2, D)$. Because of the minimality of (E, A, B, C, D), this implies that the number of rows of the matrix A_{21} should then be zero, i.e., $\text{im } [E \quad B] = X_e$. This yields (i). (Note that the reduction procedure produces a D representation for which (i) holds.) Statement (ii) can be proved by a dual argument; decompose X_d as $X_d = X_{d_1} \oplus X_{d_2}$ with $X_{d_2} = \ker E \cap \ker C$, and decompose X_e as $X_e = X_{e_1} \oplus X_{e_2}$ with $X_{e_1} = A[\ker E \cap \ker C]$. With respect to these decompositions, the matrix

$$\begin{bmatrix} sE - A & -B \\ C & D \end{bmatrix}$$

has the form

$$\begin{bmatrix} sE_{11} - A_{11} & -A_{12} & -B_1 \\ sE_{21} - A_{21} & 0 & -B_2 \\ C_1 & 0 & D \end{bmatrix}$$

where A_{12} has full row rank. Using Lemma 3.18 we have that (4.6) is strongly externally equivalent to the representation

$$\begin{bmatrix} 0 & 0 \\ 0 & 0 \\ I & 0 \\ 0 & I \end{bmatrix} \begin{bmatrix} y \\ u \end{bmatrix} = \begin{bmatrix} \sigma E_{11} - A_{11} & -A_{12} & -B_1 \\ \sigma E_{21} - A_{21} & 0 & -B_2 \\ C_1 & 0 & D \\ 0 & 0 & I \end{bmatrix} \begin{bmatrix} \xi_1 \\ \xi_2 \\ \xi_3 \end{bmatrix},$$

which is by Lemma 3.14 strongly externally equivalent to the D representation $(E_{21}, A_{21}, B_2, C_1, D)$. Because of the assumed minimality of (E, A, B, C, D), this implies that the number of columns of the matrix A_{12} should then be zero, i.e., $\ker E \cap \ker C = \{0\}$. This yields (ii). (Note that the reduction procedure produces a D representation for which (ii) holds.) Finally we prove (iii). Let $\tilde{E}$, $\tilde{A}$, $\tilde{B}$ and $\tilde{C}$ be defined as in Algorithm 3.36. Using the notation of the algorithm, we define $\tilde{D}$ as

$$\tilde{D} := D - C_2 A_{22}^{-1} B_2.$$

It can then be proven as in the proof of Lemma 3.37 that $(\tilde{E}, \tilde{A}, \tilde{B}, \tilde{C}, \tilde{D})$ is strongly externally equivalent to (E, A, B, C, D). Because of the assumed minimality of (E, A, B, C, D), the matrix A_{22} should be empty, i.e., $X_{d2} = \{0\}$ (notation of Algorithm 3.36), which implies (iii). (Note that, by Lemma 3.37, the reduction procedure produces a D representation for which (iii) holds.)

Remark 4.9 The reduction procedure in the last part of the above proof affects the size of E, but not the rank of E. The space X_{d2} consists of variables ξ that are called *nondynamic modes* in [60]. The above condition (iii) can therefore be interpreted as "absence of nondynamic modes". ◇

The following lemma is the analog of Lemma 4.5 for D representations. In the lemma π_U denotes the mapping from $W = Y \oplus U$ to U that is defined by

$$\pi_U : \begin{bmatrix} y \\ u \end{bmatrix} \mapsto u.$$

Lemma 4.10 Let $\Sigma = (\mathbf{T}, W = Y \oplus U, \mathcal{B}) \in \mathcal{L}^{p+m}$ and let (E, A, B, C, D) be a D representation of $\mathcal{B}$. Let W^0 be the DV space of Σ. Assume that the following conditions are satisfied:

(i) $[E \quad B]$ has full row rank

(ii) $\begin{bmatrix} E \\ C \end{bmatrix}$ has full column rank

(iii) $A[\ker E] \subset \operatorname{im} E$.

Then we have

(a) $C[\ker E] = Y \cap W^0$

(b) $B^{-1}[\operatorname{im} E] = \pi_U W^0$.

Proof Let (F, G, H_y, H_u) be the P representation that results from applying Algorithm 3.28 to (E, A, B, C, D). Then, according to Lemma 3.29, G has full row rank and $\ker G \cap \ker H_y \cap \ker H_u = \{0\}$. Applying Lemma 4.5 and using the notation of Algorithm 3.28 we then have

$$W^0 = \text{im} \begin{bmatrix} C_2 & D_1 - D_2 B_{22}^{-1} B_{21} \\ 0 & I \\ 0 & -B_{22}^{-1} B_{21} \end{bmatrix}.$$

It now follows immediately that (a) and (b) hold. ◇

Our strategy in this subsection will be to minimize three indices that are related to D representations: the *rank* of E, the *column defect* of E (= dim ker E), and the *row defect* of E (= codim im E). The following lemma presents lower bounds for each of the three indices.

Lemma 4.11 Let $\Sigma = (\mathbf{T}, W = Y \oplus U, \mathcal{B}) \in \mathcal{L}^{p+m}$ and let (E, A, B, C, D) be a D representation of $\mathcal{B}$; here $E, A : X_d \to X_e$. Let W^0 be the DV space of Σ. Then the following statements hold:

(i) $\text{rank } E \geq \text{ord}(\Sigma)$

(ii) $\dim \ker E \geq \dim (Y \cap W^0)$

(iii) $\text{codim im } E \geq \text{codim } (\pi_U W^0)$.

In particular we have that

(a) $\dim X_d \geq \text{ord}(\Sigma) + \dim (Y \cap W^0)$

(b) $\dim X_e \geq \text{ord}(\Sigma) + \text{codim } (\pi_U W^0)$.

Proof Since any D representation can be written in pencil form as

$$\begin{aligned} \sigma \begin{bmatrix} E & 0 \end{bmatrix} \begin{bmatrix} \xi_1 \\ \xi_2 \end{bmatrix} &= \begin{bmatrix} A & B \end{bmatrix} \begin{bmatrix} \xi_1 \\ \xi_2 \end{bmatrix} \\ \begin{bmatrix} y \\ u \end{bmatrix} &= \begin{bmatrix} C & D \\ 0 & I \end{bmatrix} \begin{bmatrix} \xi_1 \\ \xi_2 \end{bmatrix}, \end{aligned} \tag{4.7}$$

it follows immediately from Theorem 4.2 that $\text{rank } E = \text{rank } [E \quad 0] \geq \text{ord}(\Sigma)$. Next, we prove (ii) and (iii). Without loss of generality we may assume that conditions (i), (ii) and (iii) of Lemma 4.10 are satisfied since in the reduction procedures of the proof of Lemma 4.8 neither dim ker E nor codim im E is increased. It now follows from Lemma 4.10 that $\dim \ker E = \dim C[\ker E] = \dim Y \cap W^0$ and that $\text{codim im } E = \text{codim } B^{-1}[\text{ im } E] = \text{codim } (\pi_U W^0)$. But then (ii) and (iii) hold in the general case. ◇

We now present the main result of this subsection: the conditions of Lemma 4.8 together with observability at all finite modes are sufficient for minimality under strong external equivalence.

Theorem 4.12 A D representation (E, A, B, C, D) is minimal under strong external equivalence if and only if the following conditions hold:

(i) $[E \quad B]$ has full row rank

(ii) $\begin{bmatrix} E \\ C \end{bmatrix}$ has full column rank

(iii) $A[\ker E] \subset \operatorname{im} E$

(iv) $\begin{bmatrix} sE - A \\ C \end{bmatrix}$ has full column rank for all $s \in \mathbb{C}$.

Moreover, if the representation is minimal then the lower bounds in Lemma 4.11 are reached.

Proof To prove the "only if" part let us assume that (E, A, B, C, D) is minimal under strong external equivalence. The necessity of (i), (ii) and (iii) has already been proved in Lemma 4.8. We will now prove the necessity of (iv). Let V^* be the subspace of X_d that is the limit space of the iteration (2.24) applied to the pencil

$$\begin{bmatrix} sE - A \\ C \end{bmatrix},$$

that is, the iteration

$$V^0 = X_d, \qquad V^{m+1} = (A^{-1}EV^m) \cap \ker C.$$

Decompose X_d as $X_d = X_{d1} \oplus X_{d2}$ where $X_{d1} = V^*$, and decompose X_e as $X_e = X_{e1} \oplus X_{e2}$ where $X_{e1} = EV^*$. With respect to these decompositions, we may write (note that $AV^* \subset EV^*$ and that $V^* \subset \ker C$)

$$\left[\begin{array}{c} sE - A \\ \hline C \end{array}\right] = \left[\begin{array}{cc} sE_{11} - A_{11} & sE_{12} - A_{12} \\ 0 & sE_{22} - A_{22} \\ \hline 0 & C_2 \end{array}\right].$$

Then it follows from Theorem 2.31 that the matrix

$$\begin{bmatrix} sE_{22} - A_{22} \\ C_2 \end{bmatrix}$$

has full column rank for all $s \in \mathbb{C}$, whereas, by definition, E_{11} is a matrix of full row rank. Writing B with respect to the decomposition of X_e as

$$B = \begin{bmatrix} B_1 \\ B_2 \end{bmatrix},$$

we have that, by Lemma 3.18, the representation (E, A, B, C, D) is strongly externally equivalent to the representation

$$\begin{bmatrix} 0 & 0 \\ 0 & 0 \\ I & 0 \\ 0 & I \end{bmatrix} \begin{bmatrix} y \\ u \end{bmatrix} = \begin{bmatrix} \sigma E_{11} - A_{11} & \sigma E_{12} - A_{12} & -B_1 \\ 0 & \sigma E_{22} - A_{22} & -B_2 \\ 0 & C_2 & D \\ 0 & 0 & I \end{bmatrix} \begin{bmatrix} \xi_1 \\ \xi_2 \end{bmatrix}.$$

It now follows from Lemma 3.14 that (E, A, B, C, D) is strongly externally equivalent to the D representation $(E_{22}, A_{22}, B_2, C_2, D)$. Because of the assumed minimality of (E, A, B, C, D), the number of columns of the matrix $sE_{11} - A_{11}$ should then be zero, i.e., $V^* = \{0\}$. By Theorem 2.31 this implies that (iv) holds. (Note that the reduction procedure produces a D representation for which (iv) holds.) To prove the "if" part let us assume that (E, A, B, C, D) satisfies conditions (i)-(iv). We will prove that the lower bounds in Lemma 4.11 are reached. Since these lower bounds are invariant under strong external equivalence, it then follows that the representation is minimal under strong external equivalence. Because of (i), (ii) and (iii), Lemma 4.10 implies that

$$C[\,\ker E] = Y \cap W^0,$$

so that $\dim \ker E = \dim C[\,\ker E] = \dim\,(Y \cap W^0)$. By the same lemma we have that

$$B^{-1}[\,\mathrm{im}\, E] = \pi_U W^0,$$

so that $\mathrm{codim\ im}\, E = \mathrm{codim}\, B^{-1}[\,\mathrm{im}\, E] = \mathrm{codim}\,(\pi_U W^0)$. We will now show that $\mathrm{rank}\, E = \mathrm{ord}(\Sigma)$. For this we apply Algorithm 3.28 to (E, A, B, C, D); for the resulting P representation (F, G, H_y, H_u) we have (Lemma 3.29) that $\mathrm{rank}\, G = \mathrm{rank}\, E$ and that the conditions of Theorem 4.3 are satisfied. The representation (F, G, H_y, H_u) is therefore minimal under strong external equivalence, and it follows from Theorem 4.3 that $\mathrm{rank}\, G = \mathrm{ord}(\Sigma)$. We then conclude that $\mathrm{rank}\, E = \mathrm{ord}(\Sigma)$ also, and this proves the theorem. ◇

Because of Lemma 2.36, we may formulate the above minimality conditions differently as follows:

(a) $[E \quad B]$ has full row rank

(b) $\begin{bmatrix} sE - A \\ C \end{bmatrix}$ has no zeros[1] in $\mathbb{C}^e$

(c) $A[\,\ker E] \subset \mathrm{im}\, E$.

[1] Here and below "no zeros" should be understood as "no zeros of positive order".

Example 4.13 Consider the electrical circuit of Example 1.5, which is described by the D representation (1.11):

$$\begin{bmatrix} cr_2 & 0 \\ 0 & 0 \end{bmatrix} \begin{bmatrix} \dot{\xi}_1 \\ \dot{\xi}_2 \end{bmatrix} = \begin{bmatrix} -1 & r_2 \\ 1 & r_1 \end{bmatrix} \begin{bmatrix} \xi_1 \\ \xi_2 \end{bmatrix} + \begin{bmatrix} 0 \\ -1 \end{bmatrix} \boldsymbol{u}$$

$$\boldsymbol{y} = [1 \quad 0] \begin{bmatrix} \xi_1 \\ \xi_2 \end{bmatrix}.$$

Condition (ii) of Theorem 4.12 fails, so this cannot be a minimal D representation. Indeed, a minimal D representation for this system is given by the standard state space representation (1.12). In case that the resistance r_1 is absent, i.e., $r_1 = 0$, the behavior is given by

$$\boldsymbol{y} = \boldsymbol{u},$$

which is actually a minimal D representation with $X_d = X_e = \{0\}$.

Example 4.14 Consider the two interconnected electrical circuits of Example 1.6. The interconnected system is described by the D representation (1.14):

$$\begin{bmatrix} 1 & 0 \\ 0 & 1 \\ 0 & 0 \end{bmatrix} \begin{bmatrix} \dot{\boldsymbol{x}}_1 \\ \dot{\boldsymbol{x}}_2 \end{bmatrix} = \begin{bmatrix} 0 & 0 \\ 0 & 0 \\ 1 & -1 \end{bmatrix} \begin{bmatrix} \boldsymbol{x}_1 \\ \boldsymbol{x}_2 \end{bmatrix} + \begin{bmatrix} 1/c_1 & 0 \\ 0 & 1 \\ 0 & 0 \end{bmatrix} \begin{bmatrix} \boldsymbol{u}_1 \\ \boldsymbol{u}_2 \end{bmatrix}$$

$$\begin{bmatrix} \boldsymbol{y}_1 \\ \boldsymbol{y}_2 \end{bmatrix} = \begin{bmatrix} 1 & 0 \\ 0 & 1 \end{bmatrix} \begin{bmatrix} \boldsymbol{x}_1 \\ \boldsymbol{x}_2 \end{bmatrix}.$$

Condition (i) of Theorem 4.12 fails, so that this cannot be a minimal D representation. Indeed, a minimal D representation for this system is given by the D representation (1.15).

Example 4.15 Consider the Proportional-Derivative controller of Example 1.7, which is described by the D representation (1.18):

$$\begin{bmatrix} 0 & 1 \\ 0 & 0 \end{bmatrix} \begin{bmatrix} \dot{\xi}_1 \\ \dot{\xi}_2 \end{bmatrix} = \begin{bmatrix} 1 & 0 \\ 0 & 1 \end{bmatrix} \begin{bmatrix} \xi_1 \\ \xi_2 \end{bmatrix} + \begin{bmatrix} 0 \\ -1 \end{bmatrix} \boldsymbol{q}$$

$$\boldsymbol{F} = [K_d \quad K_p] \begin{bmatrix} \xi_1 \\ \xi_2 \end{bmatrix}.$$

For $K_d \neq 0$ all conditions of Theorem 4.12 are satisfied, so that this D representation is a minimal one. ◇

We conclude this subsection with three theorems referred to in Section 3.5.

Theorem 4.16 Let (F, G, H_y, H_u) be a P representation that results from applying Algorithm 3.28 to a D representation (E, A, B, C, D) that satisfies conditions (i), (ii) and (iv) of Theorem 4.12. Then (F, G, H_y, H_u) is minimal under strong external equivalence. In particular, (F, G, H_y, H_u) is minimal under strong external equivalence if (E, A, B, C, D) is minimal under strong external equivalence.

Proof By Lemma 3.29 we have that (F, G, H_y, H_u) satisfies the conditions of Theorem 4.3 and is therefore minimal under strong external equivalence. The second statement follows from the previous theorem.

Theorem 4.17 Let (E, A, B, C, D) be a D representation that results from applying Algorithm 3.30 to a P representation (F, G, H_y, H_u) that is minimal under strong external equivalence. Then (E, A, B, C, D) is also minimal under strong external equivalence.

Proof By Theorem 4.3 and Lemma 4.5 we have that rank $G = \text{ord}(\Sigma)$, $\ker G \cap \ker H = \{0\}$, and $H[\ker G] = W^0$. It now follows from Lemma 3.31 that rank $E = \text{ord}(\Sigma)$, $\dim \ker E = \dim (Y \cap W^0)$, and $\text{codim im } E = \text{codim } (Y + W^0) = \text{codim } (\pi_U W^0)$. We conclude from Theorem 4.12 that (E, A, B, C, D) is minimal under strong external equivalence.

Theorem 4.18 Let $(\tilde{E}, \tilde{A}, \tilde{B}, \tilde{C}, \tilde{D})$ be a D representation that results from applying Algorithm 3.36 to a DZ representation (E, A, B, C) that is minimal under strong external equivalence. Then $(\tilde{E}, \tilde{A}, \tilde{B}, \tilde{C}, \tilde{D})$ is also minimal under strong external equivalence.

Proof By Lemma 3.37 we have that $(\tilde{E}, \tilde{A}, \tilde{B}, \tilde{C}, \tilde{D})$ satisfies the conditions of Theorem 4.12 and is therefore minimal under strong external equivalence.

4.2.2 Results under weak external equivalence

In this subsection we consider the minimality issue of the previous subsection under weak external equivalence. As expected (see the remarks at the beginning of the section), controllability at all finite modes is an additional requirement for minimality under weak external equivalence. Together with the conditions of the previous subsection we get five necessary and sufficient conditions for minimality under weak external equivalence:

Theorem 4.19 Let $\Sigma = (\mathbf{T}, W = Y \oplus U, \mathcal{B}) \in \mathcal{L}^{p+m}$ and let (E, A, B, C, D) be a D representation of $\mathcal{B}$; here $E, A : X_d \to X_e$. Let W^0 be the DV space of Σ. Then (E, A, B, C, D) is minimal under weak external equivalence if and only if the following conditions hold:

(i) $[E \quad B]$ has full row rank

(ii) $\begin{bmatrix} E \\ C \end{bmatrix}$ has full column rank

(iii) $A[\ker E] \subset \text{im } E$

(iv) $\begin{bmatrix} sE - A \\ C \end{bmatrix}$ has full column rank for all $s \in \mathbb{C}$

(v) $[sE - A \quad B]$ has full row rank for all $s \in \mathbb{C}$.

Moreover, if the representation is minimal, then the following statements hold:

(a) rank $E = \text{C-ord}(\Sigma)$

(b) dim ker $E = \dim\,(Y \cap W^0)$

(c) codim im $E = \text{codim}\,(\pi_U W^0)$

(d) $\dim X_d = \text{C-ord}(\Sigma) + \dim\,(Y \cap W^0)$

(e) $\dim X_e = \text{C-ord}(\Sigma) + \text{codim}\,(\pi_U W^0)$.

Proof A D representation that is minimal under weak external equivalence should certainly satisfy conditions (i)-(iv) because of Theorem 4.12 and the fact that weak external equivalence is weaker than strong external equivalence. We will now prove the necessity of (v). Let T^* be the subspace of X_d that is the limit space of the following iteration:

$$T^0 = \{0\}, \quad T^{m+1} = E^{-1}[AT^m + \text{ im } B].$$

Note that T^* is the projection on X_d of the limit space of the iteration (2.29) applied to the pencil $[sE - A \quad B]$. Decompose X_d as $X_d = X_{d1} \oplus X_{d2}$ where $X_{d1} = T^*$, and decompose X_e as $X_e = X_{e1} \oplus X_{e2}$ where $X_{e1} = AT^* + \text{ im } B$. With respect to these decompositions, we may write (note that $ET^* \subset AT^* + \text{ im } B$)

$$\left[\begin{array}{c|c} sE - A & -B \\ \hline C & D \end{array}\right] = \left[\begin{array}{cc|c} sE_{11} - A_{11} & sE_{12} - A_{12} & -B_1 \\ 0 & sE_{22} - A_{22} & 0 \\ \hline C_1 & C_2 & D \end{array}\right].$$

It follows from Theorem 2.33 that the matrix $[sE_{11} - A_{11} \quad - B_1]$ has full row rank for all $s \in \mathbb{C}$. Since ker $E \subset T^*$, the matrix E_{22} has full column rank. As a result, the RC space $\mathcal{C}$ is given as (Lemma 3.16)

$$\begin{aligned} \mathcal{C} &= \begin{bmatrix} C & D \\ 0 & I \end{bmatrix} (\text{ ker } [sE - A \quad - B]) \\ &= [C_1 \quad D] (\text{ ker } [sE_{11} - A_{11} \quad - B_1]), \end{aligned}$$

and we conclude that (E, A, B, C, D) is weakly externally equivalent to the D representation $(E_{11}, A_{11}, B_1, C_1, D)$. Because of the assumed minimality of (E, A, B, C, D), the number of rows of the matrix $sE_{22} - A_{22}$ should then be zero, i.e., $AT^* + \text{ im } B = X_e$. By Theorem 2.33 this implies (v). (Note that the reduction procedure produces a D representation which satisfies (v).) To prove the "if" part let us assume that (E, A, B, C, D) satisfies conditions (i)-(v). It then follows as in the proof of Theorem 4.12 that (a)-(e) hold (use Theorem 4.7 instead of Theorem 4.3). By Lemma 4.11

we have that in general dim $X_d \geq$ C-ord(Σ) + dim $(Y \cap W^0)$ and dim $X_e \geq$ C-ord(Σ)+codim $(\pi_U W^0)$. From the fact that C-ord(Σ) and W^0 are invariant under weak external equivalence we now conclude that (F, G, H) is minimal under weak external equivalence. ◇

Because of Lemma 2.36 and Lemma 2.37, we may formulate the above minimality conditions differently as follows:

(a) $[sE - A \quad B]$ has no zeros in $\mathbb{C}^e$

(b) $\begin{bmatrix} sE - A \\ C \end{bmatrix}$ has no zeros in $\mathbb{C}^e$

(c) $A[\ker E] \subset \text{im } E$.

The minimality conditions in the above theorem coincide with results by Grimm [21]. Analogous results had been proved earlier in [60]. However, in that paper only the restricted class of D representations (E, A, B, C, D) for which $sE - A$ is invertible is considered. Also, a different notion of minimality is used: minimality is defined in terms of the *rank* of E instead of the size of E. It is proved that within the restricted class, a representation (E, A, B, C, D) is minimal in this sense if and only if the above conditions (a) and (b) hold. This is called "strong irreducibility" in [60]. It is not surprising that the above condition (c) is not required: the nondynamic modes are not related to the rank of E; see Remark 4.9. In the terminology of [60] we may summarize the result of Theorem 4.19 as follows: the representation (E, A, B, C, D) is minimal under weak external equivalence if and only if it is strongly irreducible and has no nondynamic modes.

The following three theorems show that Algorithm 3.28, Algorithm 3.30 and Algorithm 3.36 not only preserve minimality under strong external equivalence (Theorem 4.16, Theorem 4.17 and Theorem 4.18) but also under weak external equivalence.

Theorem 4.20 Let (F, G, H_y, H_u) be a P representation that results from applying Algorithm 3.28 to a D representation (E, A, B, C, D) that satisfies conditions (i), (ii), (iv) and (v) of Theorem 4.19. Then (F, G, H_y, H_u) is minimal under weak external equivalence. In particular, (F, G, H_y, H_u) is minimal under weak external equivalence if (E, A, B, C, D) is minimal under weak external equivalence.

Proof By Lemma 3.29 we have that (F, G, H_y, H_u) satisfies the conditions of Theorem 4.7 and is therefore minimal under weak external equivalence. The second statement follows from the previous theorem.

Theorem 4.21 Let (E, A, B, C, D) be a D representation that results from applying Algorithm 3.30 to a P representation (F, G, H_y, H_u) that is minimal under weak external equivalence. Then (E, A, B, C, D) is also minimal under weak external equivalence.

Proof By Theorem 4.7 and Lemma 4.5 we have that rank $G = \text{C-ord}(\Sigma)$, ker $G \cap$ ker $H = \{0\}$, and $H[\ker G] = W^0$. It now follows from Lemma 3.31 that rank $E = \text{C-ord}(\Sigma)$, dim ker $E = \dim (Y \cap W^0)$, and codim im $E = \text{codim } (Y + W^0) = \text{codim } (\pi_U W^0)$. We conclude from Theorem 4.19 that (E, A, B, C, D) is minimal under weak external equivalence.

Theorem 4.22 Let $(\tilde{E}, \tilde{A}, \tilde{B}, \tilde{C}, \tilde{D})$ be a D representation that results from applying Algorithm 3.36 to a DZ representation $(E_2 A_2 B_2 C)$ that is minimal under weak external equivalence. Then $(\tilde{E}, \tilde{A}, \tilde{B}, \tilde{C}, \tilde{D})$ is also minimal under weak external equivalence.

Proof By Lemma 3.37 we have that $(\tilde{E}, \tilde{A}, \tilde{B}, \tilde{C}, \tilde{D})$ satisfies the conditions of Theorem 4.29 and is therefore minimal under weak external equivalence.

4.3 Minimality of a DZ representation

In this section we consider the minimality issue for the class of descriptor representations with zero direct feedthrough matrix, i.e., the class of DZ representations:

$$\begin{aligned} \sigma E \xi &= A\xi + Bu \\ y &= C\xi. \end{aligned}$$

The results in this section are based on the results for P representations in Section 4.1; as in the previous section the link is provided by Algorithm 3.28. The minimality conditions in this section turn out to be slightly different from the minimality conditions that we obtained for D representations in the previous section. The main reason for this lies in the fact that a DZ representation has no matrix D in which nondynamic modes can be incorporated (see the last part of the proof of Lemma 4.8). Consequently, condition (iii) of Theorem 4.12 and Theorem 4.19 does not apply to DZ representations. However, the conditions (i) and (ii) of these theorems are still necessary: the D matrix does not play an essential role with respect to these conditions.

4.3.1 Results under strong external equivalence

When there is no D matrix in a descriptor representation, an expression for the DV space W^0 can be obtained that is slightly different from the one in Lemma 4.10:

Lemma 4.23 Let $\Sigma = (\mathbf{T}, W = Y \oplus U, \mathcal{B}) \in \mathcal{L}^{p+m}$ and let (E, A, B, C) be a DZ representation of $\mathcal{B}$. Let W^0 be the DV space of Σ. Assume that the following conditions are satisfied:

(i) $[E \quad B]$ has full row rank

(ii) $\begin{bmatrix} E \\ C \end{bmatrix}$ has full column rank.

Then we have

(a) $C[\ker E] = \pi_Y W^0$

(b) $B^{-1}[\operatorname{im} E] = U \cap W^0$.

Proof Let (F, G, H_y, H_u) be the P representation that results from applying Algorithm 3.28 to the D representation $(E, A, B, C, 0)$. Then, according to Lemma 3.29, G has full row rank and $\ker G \cap \ker H_y \cap \ker H_u = \{0\}$. Applying Lemma 4.5 and using the notation of Algorithm 3.28, we then have

$$W^0 = \operatorname{im} \begin{bmatrix} C_2 & 0 \\ 0 & I \\ -B_{22}^{-1} A_{22} & -B_{22}^{-1} B_{21} \end{bmatrix}.$$

It now follows immediately that (a) and (b) hold. ◇

The analog of Lemma 4.11 is the following lemma:

Lemma 4.24 Let $\Sigma = (\mathbf{T}, W = Y \oplus U, \mathcal{B}) \in \mathcal{L}^{p+m}$ and let (E, A, B, C) be a DZ representation of $\mathcal{B}$; here $E,\ A : X_d \to X_e$. Let W^0 be the DV space of Σ. Then the following statements hold:

(i) $\operatorname{rank} E \geq \operatorname{ord}(\Sigma)$

(ii) $\dim \ker E \geq \dim \pi_Y W^0$

(iii) $\operatorname{codim} \operatorname{im} E \geq \operatorname{codim} (U \cap W^0)$.

In particular we have that

(a) $\dim X_d \geq \operatorname{ord}(\Sigma) + \dim \pi_Y W^0$

(b) $\dim X_e \geq \operatorname{ord}(\Sigma) + \operatorname{codim} (U \cap W^0)$.

Proof See the proof of Lemma 4.11; use Lemma 4.23 instead of Lemma 4.10. ◇

We now present the main theorem of this section; the theorem shows that all conditions for D representations are necessary and sufficient for DZ representations except for the condition $A[\ker E] \subset \operatorname{im} E$, i.e., absence of nondynamic modes.

Theorem 4.25 A DZ representation (E, A, B, C) is minimal under strong external equivalence if and only if the following conditions hold:

(i) $[E \quad B]$ has full row rank

(ii) $\begin{bmatrix} E \\ C \end{bmatrix}$ has full column rank

(iii) $\begin{bmatrix} sE - A \\ C \end{bmatrix}$ has full column rank for all $s \in \mathbb{C}$.

Moreover, if the representation is minimal then the lower bounds in Lemma 4.24 are reached.

Proof See the proof of Theorem 4.12; use Lemma 4.23 instead of Lemma 4.10 and Lemma 4.24 instead of Lemma 4.11.

Example 4.26 Consider the electrical circuit of Example 1.5. For $r_1 = 0$ the behavior is given by the D representation (1.11):

$$\begin{aligned} \begin{bmatrix} cr_2 & 0 \\ 0 & 0 \end{bmatrix} \begin{bmatrix} \dot{\xi}_1 \\ \dot{\xi}_2 \end{bmatrix} &= \begin{bmatrix} -1 & r_2 \\ 1 & 0 \end{bmatrix} \begin{bmatrix} \xi_1 \\ \xi_2 \end{bmatrix} + \begin{bmatrix} 0 \\ -1 \end{bmatrix} \boldsymbol{u} \\ \boldsymbol{y} &= [1 \quad 0] \begin{bmatrix} \xi_1 \\ \xi_2 \end{bmatrix}. \end{aligned}$$

In Example 4.13 we already saw that this is not a minimal D representation. As a result, it can certainly not be a minimal DZ representation. A minimal DZ representation is instead given by

$$\begin{aligned} 0\dot{\boldsymbol{\xi}} &= \boldsymbol{\xi} + \boldsymbol{u} \\ \boldsymbol{y} &= -\boldsymbol{\xi}. \end{aligned}$$

Note that the spaces X_d and X_e corresponding to this representation are one-dimensional, whereas they are zero-dimensional for the minimal D representation $\boldsymbol{y} = \boldsymbol{u}$ that we found in Example 4.13. ◇

The following two theorems are analogues of Theorem 4.16 and Theorem 4.17, respectively.

Theorem 4.27 Consider a DZ representation (E, A, B, C) that is minimal under strong external equivalence. Let (F, G, H_y, H_u) be the P representation that results from applying Algorithm 3.28 to the D representation $(E, A, B, C, 0)$. Then (F, G, H_y, H_u) is minimal under strong external equivalence.

Proof It follows from Theorem 4.25 that $(E, A, B, C, 0)$ satisfies conditions (i), (ii) and (iv) of Theorem 4.12. The result now follows from Theorem 4.16.

Theorem 4.28 Let (E, A, B, C) be a DZ representation that results from applying Algorithm 3.32 to a P representation (F, G, H_y, H_u) that is minimal under strong external equivalence. Then (E, A, B, C) is also minimal under strong external equivalence.

Proof By Theorem 4.3 and Lemma 4.5 we have that rank $G = \text{ord}(\Sigma)$, ker $G \cap$ ker $H = \{0\}$, and $H[\text{ ker } G] = W^0$. It now follows from Lemma 3.33 that rank $E = \text{ord}(\Sigma)$, dim ker E = dim $\pi_Y W^0$, and codim im E = codim $(U \cap W^0)$. We conclude from Theorem 4.25 that (E, A, B, C) is minimal under strong external equivalence.

4.3.2 Results under weak external equivalence

In this subsection we consider the minimality issue of the previous subsection under weak external equivalence. By now the following theorem should come as no surprise:

Theorem 4.29 Let $\Sigma = (\mathbf{T}, W = Y \oplus U, \mathcal{B}) \in \mathcal{L}^{p+m}$ and let (E, A, B, C) be a DZ representation of $\mathcal{B}$; here $E,\ A : X_d \to X_e$. Let W^0 be the DV space of Σ. Then (E, A, B, C) is minimal under weak external equivalence if and only if the following conditions hold:

(i) $[E \quad B]$ has full row rank

(ii) $\begin{bmatrix} E \\ C \end{bmatrix}$ has full column rank

(iii) $\begin{bmatrix} sE - A \\ C \end{bmatrix}$ has full column rank for all $s \in \mathbb{C}$

(iv) $[sE - A \quad B]$ has full row rank for all $s \in \mathbb{C}$.

Moreover, if the representation is minimal then the following statements hold:

(a) rank $E =$ C-ord(Σ)

(b) dim ker $E =$ dim $\pi_Y W^0$

(c) codim im $E =$ codim $(U \cap W^0)$

(d) dim $X_d =$ C-ord$(\Sigma) +$ dim $\pi_Y W^0$

(e) dim $X_e =$ C-ord$(\Sigma) +$ codim $(U \cap W^0)$.

Proof See the proof of Theorem 4.19; use Lemma 4.23 instead of Lemma 4.10. ◇

The following two theorems show that Algorithm 3.28 and Algorithm 3.32 not only preserve minimality under strong external equivalence (Theorem 4.27 and Theorem 4.28) but also under weak external equivalence.

Theorem 4.30 Consider a DZ representation (E, A, B, C) that is minimal under weak external equivalence. Let (F, G, H_y, H_u) be the P representation that results from applying Algorithm 3.28 to the D representation $(E, A, B, C, 0)$. Then (F, G, H_y, H_u) is minimal under weak external equivalence.

Proof It follows from Theorem 4.29 that a DZ representation (E, A, B, C) that is minimal under weak external equivalence satisfies conditions (i), (ii), (iv) and (v) of Theorem 4.19. The result now follows from Theorem 4.20.

Theorem 4.31 Let (E, A, B, C) be a DZ representation that results from applying Algorithm 3.32 to a P representation (F, G, H_y, H_u) that is minimal under weak external equivalence. Then (E, A, B, C) is also minimal under weak external equivalence.

Proof By Theorem 4.7 and Lemma 4.5 we have that rank $G =$ C-ord(Σ), ker $G \cap$ ker $H = \{0\}$ and $H[\ker G] = W^0$. It now follows from Lemma 3.33 that rank $E =$ C-ord(Σ), dim ker $E =$ dim $\pi_Y W^0$, and codim im $E =$ codim $(U \cap W^0)$. We conclude from Theorem 4.29 that (E, A, B, C) is minimal under weak external equivalence.

4.4 Minimality of a DP representation

In this section we investigate the minimality issue for a DP representation

$$\sigma K \xi = L \xi + M w.$$

As before, K and L are mappings from X to Z. The representation will be called *minimal* if both Z and X have least dimension in the class of equivalent DP representations. A minimal DP representation thus involves a minimum number of internal variables *and* a minimum number of equations. As mentioned at the beginning of Section 3.5, we may interpret a DP representation (K, L, M) as the DZ representation (E, A, B, C) where $E = K$, $A = L$ and $B = M$, and $C : X \to \{0\}$ is the zero mapping. Results on minimality for DP representations are therefore immediate from the previous section. In the following two subsections we present the results under strong external equivalence and weak external equivalence.

4.4.1 Results under strong external equivalence

The following theorem is a corollary of Theorem 4.25.

Theorem 4.32 Let $\Sigma = (\mathbf{T}, W, \mathcal{B}) \in \mathcal{L}^q$ and let (K, L, M) be a DP representation of $\mathcal{B}$; here K, $L : X \to Z$. Let W^0 be the DV space of Σ. Then (K, L, M) is minimal under strong external equivalence if and only if the following conditions hold:

(i) K has full column rank

(ii) $[K \quad M]$ has full row rank

(iii) $sK - L$ has full column rank for all $s \in \mathbb{C}$.

Moreover, if the representation is minimal then the following statements hold:

(a) $\dim X = \text{ord}(\Sigma)$

(b) $\dim Z = \text{ord}(\Sigma) + \text{codim } W^0$.

◇

The minimality issue of this subsection has been considered for $\mathbf{T} = \mathbb{Z}$ by Willems in [68]. The conditions that are derived in [68] differ from ours: there is the additional requirement that $[L \quad M]$ should have full row rank. Briefly, this is due to the fact that for $\mathbf{T} = \mathbb{Z}$ we have $\sigma\mathcal{B} = \mathcal{B}$ rather than $\sigma\mathcal{B} \subset \mathcal{B}$; see also the remarks in [66].

4.4.2 Results under weak external equivalence

The following theorem is a corollary of Theorem 4.29. The minimality conditions in the theorem coincide with results in [2, 3].

Theorem 4.33 Let $\Sigma = (\mathbf{T}, W, \mathcal{B}) \in \mathcal{L}^q$ and let (K, L, M) be a DP representation of $\mathcal{B}$; here K, $L : X \to Z$. Let W^0 be the DV space of Σ. Then (K, L, M) is minimal under weak external equivalence if and only if the following conditions hold:

(i) K has full column rank

(ii) $[K \quad M]$ has full row rank

(iii) $sK - L$ has full column rank for all $s \in \mathbb{C}$

(iv) $[sK - L \quad M]$ has full row rank for all $s \in \mathbb{C}$.

Moreover, if the representation is minimal then

(a) $\dim X = \text{C-ord}(\Sigma)$

(b) $\dim Z = \text{C-ord}(\Sigma) + \operatorname{codim} W^0$.

◇

The above theorem can also be derived on the basis of the duality that exists under weak external equivalence between P representations and DP representations. Using Lemma 3.23 and Lemma 3.26 we can then obtain the theorem as a corollary of Theorem 4.7.

4.5 Transformation groups

In this section it is our aim to derive an operational form of equivalence for each of the first-order representations that we considered in the previous sections. Of course, a change of basis in internal/equation space should be included in the operational form. At a general level the operational forms will be quite involved, since operations, according to which redundant variables can be eliminated, should also be included. For representations without redundancy, i.e., minimal representations, the operations are much less involved and even form a group which will be referred to as the "transformation group". For the (minimal) first-order representations that we consider here we will show that the transformation group is entirely made up of "change of basis in internal/equation space". There is one exception, however, as we will see shortly.

In the past, many concepts of equivalence have been defined for descriptor representations. We mention "restricted system equivalence", introduced by Rosenbrock [52] for DZ representations as the analog of "Kalman equivalence" as defined for standard state space representations. Two DZ representations (E, A, B, C) and $(\tilde{E}, \tilde{A}, \tilde{B}, \tilde{C})$ are defined to be restricted system equivalent if they are related by a change of basis in descriptor/equation space: there exist nonsingular matrices M and N such that

$$\begin{bmatrix} M & 0 \\ 0 & I \end{bmatrix} \begin{bmatrix} sE - A & -B \\ C & 0 \end{bmatrix} \begin{bmatrix} N & 0 \\ 0 & I \end{bmatrix} = \begin{bmatrix} s\tilde{E} - \tilde{A} & -\tilde{B} \\ \tilde{C} & 0 \end{bmatrix}.$$

For descriptor representations with direct feedthrough matrix the above type of equivalence was not considered useful in [60]: the non-dynamic modes are not treated in a satisfactory way. For this reason the concept of "strong equivalence" was introduced in [60] as a modified version of restricted system equivalence. It involves so-called "operations of strong equivalence" that are defined as follows: two D representations (E, A, B, C, D) and $(\tilde{E}, \tilde{A}, \tilde{B}, \tilde{C}, \tilde{D})$ are defined to be related by operations of strong equivalence if there exist matrices M, N, R and Q with M and N invertible such that

$$\begin{bmatrix} M & 0 \\ R & I \end{bmatrix} \begin{bmatrix} sE - A & -B \\ C & D \end{bmatrix} \begin{bmatrix} N & Q \\ 0 & I \end{bmatrix} = \begin{bmatrix} s\tilde{E} - \tilde{A} & -\tilde{B} \\ \tilde{C} & \tilde{D} \end{bmatrix}.$$

Note that for a D representation

$$\sigma E \boldsymbol{\xi} = A\boldsymbol{\xi} + B\boldsymbol{u} \tag{4.8}$$

$$\boldsymbol{y} = C\boldsymbol{\xi} + D\boldsymbol{u}, \tag{4.9}$$

the algebraic equations that are implicit in (4.8) are allowed to be added to (4.9) under "operations of strong equivalence", but not under "restricted system equivalence".

In the following subsections we derive transformation groups under both strong and weak external equivalence for minimal P representations, D representations, DZ representations and DP representations.

4.5.1 Results for minimal P representations

The results of the next theorem are essentially present in [65, Theorem 7.1]; see the remarks in [29]. Our proof, however, is more elaborate.

Theorem 4.34 Let (F, G, H) and $(\tilde{F}, \tilde{G}, \tilde{H})$ be P representations that are minimal under strong external equivalence. Then the representations are strongly externally equivalent if and only if there exist nonsingular matrices S and T such that

$$\begin{bmatrix} S & 0 \\ 0 & I \end{bmatrix} \begin{bmatrix} sG - F \\ H \end{bmatrix} T = \begin{bmatrix} s\tilde{G} - \tilde{F} \\ \tilde{H} \end{bmatrix}. \tag{4.10}$$

Proof The "if" part follows from Lemma 3.18. To prove the "only if" part let (F, G, H) and $(\tilde{F}, \tilde{G}, \tilde{H})$ be P representations that are strongly externally equivalent and minimal under strong external equivalence. Let $\mathcal{B}$ be the behavior that corresponds to the representations; let $\mathcal{W}_0$, $\mathcal{C}$ and $\mathcal{W}^0$ be the corresponding RB space, RC space and DV space, respectively. Because of minimality (Theorem 4.3), both G and $\tilde{G}$ have full row rank. Without loss of generality we may assume that $G = \tilde{G} = [I \quad 0]$, since the (group) operations that bring G and $\tilde{G}$ in this form are included in (4.10) (note that G and $\tilde{G}$ have the same rank and size because of minimality). Accordingly write $F = [F_1 \quad F_2]$, $H = [H_1 \quad H_2]$ and similarly for $\tilde{F}$ and

$\tilde{H}$. Decompose W as $W = W_1 \oplus W_2$ with $W_2 = W^0$. It is sufficient to prove the theorem with respect to this decomposition since the equality

$$\begin{bmatrix} S & 0 \\ 0 & M \end{bmatrix} \begin{bmatrix} sG - F \\ H \end{bmatrix} T = \begin{bmatrix} I & 0 \\ 0 & M \end{bmatrix} \begin{bmatrix} s\tilde{G} - \tilde{F} \\ \tilde{H} \end{bmatrix},$$

where M is a nonsingular matrix, implies (4.10). Write

$$H_1 = \begin{bmatrix} H_{11} \\ H_{21} \end{bmatrix}, H_2 = \begin{bmatrix} H_{12} \\ H_{22} \end{bmatrix}, \tilde{H}_1 = \begin{bmatrix} \tilde{H}_{11} \\ \tilde{H}_{21} \end{bmatrix}, \tilde{H}_2 = \begin{bmatrix} \tilde{H}_{12} \\ \tilde{H}_{22} \end{bmatrix}$$

with respect to the above decomposition of W. The P representation (F, G, H) then becomes

$$\begin{aligned} \sigma\xi_1 &= F_1\xi_1 + F_2\xi_2 \\ w_1 &= H_{11}\xi_1 + H_{12}\xi_2 \\ w_2 &= H_{21}\xi_1 + H_{22}\xi_2 \end{aligned} \tag{4.11}$$

and similarly for $(\tilde{F}, \tilde{G}, \tilde{H})$. Because of minimality, dim ker G = dim ker $\tilde{G}$ = dim W^0 and $H[\ker G] = \tilde{H}[\ker \tilde{G}] = W^0$ (Theorem 4.3 and Lemma 4.5). As a result the matrices H_{22} and $\tilde{H}_{22}$ are nonsingular. The representation (4.11) can then be written as

$$\begin{aligned} \sigma\xi_1 &= A\xi_1 + Bw_2 \\ w_1 &= C\xi_1 + Dw_2 \end{aligned}$$

where A, B, C and D are defined by

$$\left[\begin{array}{c|c} sI - A & -B \\ \hline C & D \end{array}\right] := \left[\begin{array}{cc} sI - F_1 & -F_2 \\ \hline H_{11} & H_{12} \end{array}\right] \left[\begin{array}{c|c} I & 0 \\ -H_{22}^{-1}H_{21} & H_{22}^{-1} \end{array}\right]. \tag{4.12}$$

By Theorem 4.3 the matrix $\begin{bmatrix} sG^T - F^T & H^T \end{bmatrix}^T$ has full column rank for all $s \in \mathbb{C}$. Because of (4.12), it follows that $\begin{bmatrix} sI - A^T & C^T \end{bmatrix}^T$ also has full column rank for all $s \in \mathbb{C}$. From Lemma 3.16 we derive the following expression for $\mathcal{W}_0$ and $\mathcal{C}$:

$$\begin{aligned} \mathcal{W}_0 &= \{\begin{pmatrix} w_1(\lambda) \\ w_2(\lambda) \end{pmatrix} \in \lambda^{-1}W_\infty(\lambda) \mid \exists \xi_0 \in Z \text{ such that} \\ &\quad w_1(\lambda) = C(\lambda I - A)^{-1}\xi_0 + (C(\lambda I - A)^{-1}B + D)w_2(\lambda)\} \end{aligned} \tag{4.13}$$

$$\mathcal{C} = \text{im} \begin{bmatrix} C(\lambda I - A)^{-1}B + D \\ I \end{bmatrix}. \tag{4.14}$$

Introducing $\tilde{A}$, $\tilde{B}$, $\tilde{C}$ and $\tilde{D}$ in a similar way, we conclude from the strong external equivalence of (F, G, H) and $(\tilde{F}, \tilde{G}, \tilde{H})$ that $\mathcal{W}_0$ is also given by

$$\mathcal{W}_0 = \{\begin{pmatrix} w_1(\lambda) \\ w_2(\lambda) \end{pmatrix} \in \lambda^{-1} W_\infty(\lambda) \mid \exists \xi_0 \in Z \text{ such that}$$

$$w_1(\lambda) = \tilde{C}(\lambda I - \tilde{A})^{-1}\xi_0 + (\tilde{C}\left(\lambda I - \tilde{A}\right)^{-1}\tilde{B} + \tilde{D})w_2(\lambda)\} \tag{4.15}$$

and that $\mathcal{C}$ is also given by

$$\mathcal{C} = \text{ im} \begin{bmatrix} \tilde{C}\left(\lambda I - \tilde{A}\right)^{-1}\tilde{B} + \tilde{D} \\ I \end{bmatrix}. \tag{4.16}$$

We conclude from (4.14) and (4.16) that

$$C(\lambda I - A)^{-1}B + D = \tilde{C}\left(\lambda I - \tilde{A}\right)^{-1}\tilde{B} + \tilde{D} \tag{4.17}$$

and, using this equality in (4.13) and (4.15), that there exists a square matrix N such that

$$C(\lambda I - A)^{-1}N = \tilde{C}\left(\lambda I - \tilde{A}\right)^{-1}.$$

We now use Lemma 2.34 to conclude that N has full column rank and hence is nonsingular. Applying the standard state space isomorphism theorem (see, for instance, [26, Theorem 6.2-4]) to the standard state space representations (A, N, C) and $(\tilde{A}, I, \tilde{C})$ we get

$$CN = \tilde{C}, \quad N^{-1}AN = \tilde{A}.$$

From (4.17) it follows that $D = \tilde{D}$ and

$$\begin{aligned} C(\lambda I - A)^{-1}B &= \tilde{C}\left(\lambda I - \tilde{A}\right)^{-1}\tilde{B} \\ &= C(\lambda I - A)^{-1}N\tilde{B}. \end{aligned}$$

Again using Lemma 2.34 we conclude that $NB = \tilde{B}$. We now have that

$$\begin{bmatrix} N^{-1} & 0 & 0 \\ 0 & I & 0 \\ 0 & 0 & I \end{bmatrix} \begin{bmatrix} sI - A & -B \\ C & D \\ 0 & I \end{bmatrix} \begin{bmatrix} N & 0 \\ 0 & I \end{bmatrix} = \begin{bmatrix} sI - \tilde{A} & -\tilde{B} \\ \tilde{C} & \tilde{D} \\ 0 & I \end{bmatrix}.$$

From (4.12) it now follows that (4.10) holds with $S = N^{-1}$ and

$$T = \begin{bmatrix} I & 0 \\ -H_{22}{}^{-1}H_{21} & H_{22}{}^{-1} \end{bmatrix} \begin{bmatrix} N & 0 \\ 0 & I \end{bmatrix} \begin{bmatrix} I & 0 \\ -\tilde{H}_{22}^{-1}\tilde{H}_{21} & \tilde{H}_{22}^{-1} \end{bmatrix}^{-1}.$$

◇

The class of P representations that are minimal under *weak* external equivalence is smaller than the class of P representations that we considered above. The next theorem shows that P representations in this class that are weakly externally equivalent are related by the transformation group of Theorem 4.34.

Theorem 4.35 Let (F, G, H) and $(\tilde{F}, \tilde{G}, \tilde{H})$ be P representations that are minimal under weak external equivalence. Then the representations are weakly externally equivalent if and only if there exist nonsingular matrices S and T such that

$$\begin{bmatrix} S & 0 \\ 0 & I \end{bmatrix} \begin{bmatrix} sG - F \\ H \end{bmatrix} T = \begin{bmatrix} s\tilde{G} - \tilde{F} \\ \tilde{H} \end{bmatrix}. \tag{4.18}$$

Proof The "if" part follows from the fact that the relation (4.18) implies that $H[\ker (sG - F)] = \tilde{H}[\ker (s\tilde{G} - \tilde{F})]$ (use Lemma 3.16). To prove the "only if" part let (F, G, H) and $(\tilde{F}, \tilde{G}, \tilde{H})$ be P representations that are weakly externally equivalent and minimal under weak external equivalence. Let $\mathcal{C}$ and $\mathcal{W}^0$ be the corresponding RC space and DV space, respectively. Let the matrices A, B, C, D, $\tilde{A}$, $\tilde{B}$, $\tilde{C}$ and $\tilde{D}$ be defined as in the proof of Theorem 4.34. We conclude from the fact that $sG - F$ has full row rank for all $s \in \mathbb{C}$ (Theorem 4.7), that the matrix $[sI - A \quad B]$ has full row rank for all $s \in \mathbb{C}$ (use (4.12)). A similar statement holds for $\left[sI - \tilde{A} \quad \tilde{B}\right]$. Furthermore, as in the proof of Theorem 4.34, it follows from the fact that $\mathcal{C}$ is invariant under weak external equivalence that

$$C(\lambda I - A)^{-1}B + D = \tilde{C}\left(\lambda I - \tilde{A}\right)^{-1}\tilde{B} + \tilde{D}.$$

This implies that $D = \tilde{D}$. Applying the standard state space isomorphism theorem (see, for instance, [26, Theorem 6.2-4]) to the standard state space representations (A, B, C) and $(\tilde{A}, \tilde{B}, \tilde{C})$, we conclude that there exists a nonsingular matrix N such that

$$CN = \tilde{C}, \quad N^{-1}AN = \tilde{A}, \text{ and } N^{-1}B = \tilde{B}.$$

The desired result now follows as in the proof of Theorem 4.34.

4.5.2 Results for minimal D representations

The results in this subsection are based on the results for P representations in the previous subsection; the link is provided by Algorithm 3.28. The next theorem shows that the transformation group under strong external equivalence for minimal D representations consists of more than change of basis in descriptor/equation space: minimal D representations are related by "operations of strong external equivalence" as introduced in [60]; see the remarks at the beginning of the section.

Theorem 4.36 Let (E, A, B, C, D) and $(\tilde{E}, \tilde{A}, \tilde{B}, \tilde{C}, \tilde{D})$ be D representations that are minimal under strong external equivalence. Then the representations are strongly externally equivalent if and only if there exist matrices M, N, R and Q with M and N nonsingular such that

$$\begin{bmatrix} M & 0 \\ R & I \end{bmatrix} \begin{bmatrix} sE - A & -B \\ C & D \end{bmatrix} \begin{bmatrix} N & Q \\ 0 & I \end{bmatrix} = \begin{bmatrix} s\tilde{E} - \tilde{A} & -\tilde{B} \\ \tilde{C} & \tilde{D} \end{bmatrix}. \tag{4.19}$$

Proof The "if" part follows immediately from Lemma 3.18. To prove the "only if" part let (E, A, B, C, D) and $(\tilde{E}, \tilde{A}, \tilde{B}, \tilde{C}, \tilde{D})$ be D representations that are strongly externally equivalent and minimal under strong external equivalence. Let $\mathcal{B}$ be the behavior that corresponds to the representations, and let W^0 be the corresponding DV space. Without loss of generality we may assume that both E and $\tilde{E}$ are given as

$$\begin{bmatrix} I & 0 \\ 0 & 0 \end{bmatrix} \tag{4.20}$$

since the (group) operations that bring E and $\tilde{E}$ in this form are included in (4.19) (note that E and $\tilde{E}$ have the same rank and size because of minimality). Accordingly write

$$A = \begin{bmatrix} A_{11} & A_{12} \\ A_{21} & A_{22} \end{bmatrix}, B = \begin{bmatrix} B_1 \\ B_2 \end{bmatrix}, C = \begin{bmatrix} C_1 & C_2 \end{bmatrix},$$

and similarly for $\tilde{A}$, $\tilde{B}$ and $\tilde{C}$. Decompose U as $U = U_1 \oplus U_2$ with $U_2 = \pi_U W^0$. It is sufficient to prove the theorem with respect to this decomposition since the equality

$$\begin{bmatrix} M & 0 \\ R & I \end{bmatrix} \begin{bmatrix} sE - A & -B \\ C & D \end{bmatrix} \begin{bmatrix} N & Q \\ 0 & L \end{bmatrix} = \begin{bmatrix} s\tilde{E} - \tilde{A} & -\tilde{B} \\ \tilde{C} & \tilde{D} \end{bmatrix} \begin{bmatrix} I & 0 \\ 0 & L \end{bmatrix},$$

where L is a nonsingular matrix, implies (4.19). With respect to the above decomposition of U, we write

$$B_1 = [B_{11} \quad B_{12}], B_2 = [B_{21} \quad B_{22}], D = [D_1 \quad D_2],$$

and similarly for $\tilde{B}_1$, $\tilde{B}_2$ and $\tilde{D}$. Because of minimality (Lemma 4.10 and Theorem 4.12), we have that

$$\text{codim im } E = \text{codim im } \tilde{E} = \text{codim } \pi_U W^0$$

and

$$B^{-1}[\text{ im } E] = \tilde{B}^{-1}[\text{ im } \tilde{E}] = \pi_U W^0.$$

As a result, the matrices B_{22} and $\tilde{B}_{22}$ are nonsingular. Let (F, G, H_y, H_u) and $(\tilde{F}, \tilde{G}, \tilde{H}_y, \tilde{H}_u)$ be the P representations that result from applying Algorithm 3.28 to (E, A, B, C, D) and $(\tilde{E}, \tilde{A}, \tilde{B}, \tilde{C}, \tilde{D})$, respectively. Then (F, G, H_y, H_u) and $(\tilde{F}, \tilde{G}, \tilde{H}_y, \tilde{H}_u)$ are strongly externally equivalent and minimal (Theorem 4.16). As a result of (3.37) and Theorem 4.34, there exist nonsingular matrices S and T such that

$$\begin{bmatrix} S & 0 & 0 & 0 \\ 0 & I & 0 & 0 \\ 0 & 0 & I & 0 \\ 0 & 0 & 0 & I \end{bmatrix} \begin{bmatrix} I & -B_{12}B_{22}^{-1} & 0 & 0 \\ 0 & D_2 B_{22}^{-1} & I & 0 \\ 0 & 0 & 0 & I \\ 0 & B_{22}^{-1} & 0 & 0 \end{bmatrix} \begin{bmatrix} sI - A_{11} & -A_{12} & -B_{11} \\ -A_{21} & -A_{22} & -B_{21} \\ C_1 & C_2 & D_1 \\ 0 & 0 & I \end{bmatrix} \times$$

$$\times \begin{bmatrix} T_{11} & T_{12} & T_{13} \\ T_{21} & T_{22} & T_{23} \\ T_{31} & T_{32} & T_{33} \end{bmatrix} = \begin{bmatrix} I & -\tilde{B}_{12}\tilde{B}_{22}^{-1} & 0 & 0 \\ 0 & \tilde{D}_2 \tilde{B}_{22}^{-1} & I & 0 \\ 0 & 0 & 0 & I \\ 0 & \tilde{B}_{22}^{-1} & 0 & 0 \end{bmatrix} \begin{bmatrix} sI - \tilde{A}_{11} & -\tilde{A}_{12} & -\tilde{B}_{11} \\ -\tilde{A}_{21} & -\tilde{A}_{22} & -\tilde{B}_{21} \\ \tilde{C}_1 & \tilde{C}_2 & \tilde{D}_1 \\ 0 & 0 & I \end{bmatrix}.$$

Comparing elements of the first and third rows of both sides of the above equation, we conclude that $T_{11} = S^{-1}$, $T_{12} = 0$, $T_{13} = 0$, $T_{31} = 0$, $T_{32} = 0$, and $T_{33} = I$. From the fact that T is nonsingular it now follows that T_{22} is nonsingular. From the above equation it also follows that

$$\begin{bmatrix} S & 0 & 0 \\ 0 & I & 0 \\ 0 & 0 & I \end{bmatrix} \begin{bmatrix} I & -B_{12}B_{22}^{-1} & 0 \\ 0 & D_2 B_{22}^{-1} & I \\ 0 & B_{22}^{-1} & 0 \end{bmatrix} \begin{bmatrix} sI - A_{11} & -A_{12} & -B_{11} & -B_{12} \\ -A_{21} & -A_{22} & -B_{21} & -B_{22} \\ C_1 & C_2 & D_1 & D_2 \end{bmatrix} \times$$

$$\times \begin{bmatrix} S^{-1} & 0 & 0 & 0 \\ T_{21} & T_{22} & T_{23} & 0 \\ 0 & 0 & I & 0 \\ 0 & 0 & 0 & I \end{bmatrix} = \begin{bmatrix} I & -\tilde{B}_{12}\tilde{B}_{22}^{-1} & 0 \\ 0 & \tilde{D}_2 \tilde{B}_{22}^{-1} & I \\ 0 & \tilde{B}_{22}^{-1} & 0 \end{bmatrix} \times$$

$$\times \begin{bmatrix} sI - \tilde{A}_{11} & -\tilde{A}_{12} & -\tilde{B}_{11} & -\tilde{B}_{12} \\ -\tilde{A}_{21} & -\tilde{A}_{22} & -\tilde{B}_{21} & -\tilde{B}_{22} \\ \tilde{C}_1 & \tilde{C}_2 & \tilde{D}_1 & \tilde{D}_2 \end{bmatrix}. \tag{4.21}$$

Multiplying this equation from the left by

$$\begin{bmatrix} I & -\tilde{B}_{12}\tilde{B}_{22}^{-1} & 0 \\ 0 & \tilde{D}_2 \tilde{B}_{22}^{-1} & I \\ 0 & \tilde{B}_{22}^{-1} & 0 \end{bmatrix}^{-1} = \begin{bmatrix} I & 0 & \tilde{B}_{12} \\ 0 & 0 & \tilde{B}_{22} \\ 0 & I & -\tilde{D}_2 \end{bmatrix}$$

yields

$$\begin{bmatrix} S & -SB_{12}B_{22}^{-1} + \tilde{B}_{12}B_{22}^{-1} & 0 \\ 0 & \tilde{B}_{22}B_{22}^{-1} & 0 \\ 0 & D_2 B_{22}^{-1} - \tilde{D}_2 B_{22}^{-1} & I \end{bmatrix} \begin{bmatrix} sI - A_{11} & -A_{12} & -B_{11} & -B_{12} \\ -A_{21} & -A_{22} & -B_{21} & -B_{22} \\ C_1 & C_2 & D_1 & D_2 \end{bmatrix} \times$$

$$\times \begin{bmatrix} S^{-1} & 0 & 0 & 0 \\ T_{21} & T_{22} & T_{23} & 0 \\ 0 & 0 & I & 0 \\ 0 & 0 & 0 & I \end{bmatrix} = \begin{bmatrix} sI - \tilde{A}_{11} & -\tilde{A}_{12} & -\tilde{B}_{11} & -\tilde{B}_{12} \\ -\tilde{A}_{21} & -\tilde{A}_{22} & -\tilde{B}_{21} & -\tilde{B}_{22} \\ \tilde{C}_1 & \tilde{C}_2 & \tilde{D}_1 & \tilde{D}_2 \end{bmatrix}.$$

The theorem is now proven by taking

$$M = \begin{bmatrix} S & -SB_{12}B_{22}^{-1} + \tilde{B}_{12}B_{22}^{-1} \\ 0 & \tilde{B}_{22}B_{22}^{-1} \end{bmatrix}, \quad N = \begin{bmatrix} S^{-1} & 0 \\ T_{21} & T_{22} \end{bmatrix},$$

$$R = \begin{bmatrix} 0 & D_2 B_{22}^{-1} - \tilde{D}_2 B_{22}^{-1} \end{bmatrix}, \quad Q = \begin{bmatrix} 0 & 0 \\ T_{23} & 0 \end{bmatrix}.$$

◇

As in the previous subsection the situation with respect to *weak* external equivalence resembles the situation with respect to strong external equivalence. The next theorem coincides with a result in [21].

Theorem 4.37 Let (E, A, B, C, D) and $(\tilde{E}, \tilde{A}, \tilde{B}, \tilde{C}, \tilde{D})$ be D representations that are minimal under weak external equivalence. Then the representations are weakly externally equivalent if and only if there exist matrices M, N, R and Q with M and N nonsingular such that

$$\begin{bmatrix} M & 0 \\ R & I \end{bmatrix} \begin{bmatrix} sE - A & -B \\ C & D \end{bmatrix} \begin{bmatrix} N & Q \\ 0 & I \end{bmatrix} = \begin{bmatrix} s\tilde{E} - \tilde{A} & -\tilde{B} \\ \tilde{C} & \tilde{D} \end{bmatrix}.$$

Proof See the proof of Theorem 4.36; use Theorem 4.20 instead of Theorem 4.16 and Theorem 4.35 instead of Theorem 4.34.

4.5.3 Results for minimal DZ representations

The results in this subsection are based on the results for P representations in Subsection 4.5.1; as in the previous subsection the link is provided by Algorithm 3.28. The transformation group for minimal DZ representations turns out to be different from the transformation group for minimal D representations that we obtained in the previous subsection. This is due to the fact that some of the operations in the transformation group of the previous subsection create a nonzero D matrix in a minimal DZ representation. The transformation group in this subsection is therefore more restricted. The next theorem shows that DZ representations are related by "restricted equivalence" as introduced in [52]; see the remarks at the beginning of the section.

Theorem 4.38 Let (E, A, B, C) and $(\tilde{E}, \tilde{A}, \tilde{B}, \tilde{C})$ be DZ representations that are minimal under strong external equivalence. Then the representations are strongly externally equivalent if and only if there exist nonsingular matrices M and N such that

$$\begin{bmatrix} M & 0 \\ 0 & I \end{bmatrix} \begin{bmatrix} sE - A & -B \\ C & 0 \end{bmatrix} \begin{bmatrix} N & 0 \\ 0 & I \end{bmatrix} = \begin{bmatrix} s\tilde{E} - \tilde{A} & -\tilde{B} \\ \tilde{C} & 0 \end{bmatrix}.$$

Proof See the proof of Theorem 4.36; decompose U as $U = U_1 \oplus U_2$ with $U_2 = U \cap W^0$ instead of $U_2 = \pi_U W^0$; use Lemma 4.23 instead of Lemma 4.10, Theorem 4.25 instead of Theorem 4.12 and Theorem 4.27 instead of Theorem 4.16. In the notation of the proof of Theorem 4.36 we then have that $D_1 = \tilde{D}_1 = 0$ and $D_2 = \tilde{D}_2 = 0$. Comparing the (2,3) elements on both sides of equation (4.21), we conclude that $C_2T_{23} = 0$, which implies that $T_{23} = 0$ since C_2 has full column rank (condition (ii) in Theorem 4.25). Consequently, $R = 0$ and $Q = 0$ in the notation of Theorem 4.36, which yields the desired result. ◇

As in the previous subsection the transformation group under weak external equivalence is the same as the transformation group under strong external equivalence; the proof of the following theorem is analogous to earlier proofs.

Theorem 4.39 Let (E, A, B, C) and $(\tilde{E}, \tilde{A}, \tilde{B}, \tilde{C})$ be DZ representations that are minimal under weak external equivalence. Then the representations are weakly externally equivalent if and only if there exist nonsingular matrices M and N such that

$$\begin{bmatrix} M & 0 \\ 0 & I \end{bmatrix} \begin{bmatrix} sE - A & -B \\ C & 0 \end{bmatrix} \begin{bmatrix} N & 0 \\ 0 & I \end{bmatrix} = \begin{bmatrix} s\tilde{E} - \tilde{A} & -\tilde{B} \\ \tilde{C} & 0 \end{bmatrix}.$$

4.5.4 Results for minimal DP representations

In this subsection we proceed as in Section 4.4: we interpret a DP representation (K, L, M) as the DZ representation (E, A, B, C) where $E = K$, $A = L$ and $B = M$, and $C : X \to \{0\}$ is the zero mapping. Results are then immediate from the previous subsection. The following result under strong external equivalence is a corollary of Theorem 4.38:

Theorem 4.40 Let (K, L, M) and $(\tilde{K}, \tilde{L}, \tilde{M})$ be DP representations that are minimal under strong external equivalence. Then the representations are strongly externally equivalent if and only if there exist nonsingular matrices S and T such that

$$T \begin{bmatrix} sK - L & M \end{bmatrix} \begin{bmatrix} S & 0 \\ 0 & I \end{bmatrix} = \begin{bmatrix} s\tilde{K} - \tilde{L} & \tilde{M} \end{bmatrix}.$$

◇

The transformation group under strong external equivalence for minimal DP representations has been derived for the case $\mathbf{T} = \mathbb{Z}$ in [68, Theorem VII.9]. The transformation group coincides with the transformation group in the above theorem.

The next theorem can be derived either as a corollary of Theorem 4.39 or, using the duality between the pencil form and the dual pencil form of Lemma 3.26, as a corollary of Theorem 4.35.

Theorem 4.41 Let (K, L, M) and $(\tilde{K}, \tilde{L}, \tilde{M})$ be DP representations that are minimal under weak external equivalence. Then the representations are weakly externally equivalent if and only if there exist nonsingular matrices S and T such that

$$T\begin{bmatrix} sK - L & M \end{bmatrix}\begin{bmatrix} S & 0 \\ 0 & I \end{bmatrix} = \begin{bmatrix} s\tilde{K} - \tilde{L} & \tilde{M} \end{bmatrix}.$$

5

Realization in minimal first-order form

When a first-order representation is equivalent to a representation in AR(MA) form, it is usually called a *realization* of the AR(MA) representation. In this chapter we study methods for obtaining minimal first-order realizations of AR(MA) representations under strong/weak external equivalence. Without the requirement of minimality, the issue is easily solved by methods which involve just a renaming of variables. An AR representation is then most naturally realized in dual pencil form: realization "per row" of the corresponding polynomial matrix $R(s) = R_k s^k + R_{k-1}s^{k-1} + \cdots + R_0$ leads to the DP representation (see also [3, Section 1.2])

$$\sigma \begin{bmatrix} I & & \\ & \ddots & \\ & & I \\ & & 0 \end{bmatrix} \xi = \begin{bmatrix} 0 & & \\ I & & \\ & \ddots & \\ & & I \end{bmatrix} \xi + \begin{bmatrix} R_0 \\ \vdots \\ R_k \end{bmatrix} w.$$

In an analogous way an MA representation given by a polynomial matrix $Q(s) = Q_k s^k + Q_{k-1}s^{k-1} + \cdots + Q_0$ is most naturally realized "per column" in pencil form as (see also [3, Section 1.3])

$$\begin{aligned} \sigma \begin{bmatrix} I & & & \\ & \ddots & & \\ & & I & 0 \end{bmatrix} \xi &= \begin{bmatrix} 0 & I & & \\ & & \ddots & \\ & & & I \end{bmatrix} \xi \\ w &= \begin{bmatrix} Q_0 & \cdots & Q_k \end{bmatrix} \xi. \end{aligned}$$

When the matrix $R(s)$ ($Q(s)$) consists of only one row (column), the above realization method yields a minimal DP representation (P representation), as can be seen by checking the dimensions. If, however, $R(s)$ ($Q(s)$) has multiple rows (columns), then the above realization will *not* be minimal. In this chapter we will present a method that yields minimal realizations independently of the size of $R(s)$ or $Q(s)$.

From the above one could get the impression that AR representations are naturally linked with DP representations, whereas MA representations have a natural connection with the pencil form. However, in the sequel the situation will be exactly the converse. Indeed, when looking for basis-free methods to realize AR representations in minimal first-order form, we find that the pencil form is most suitable. To get this clear, let us consider an AR representation for which $R(s)$ is nonsingular. It is important to notice that such representations are nontrivial in the context of strong external equivalence, whereas they are trivial (equivalent to the trivial system with only the zero trajectory) under weak external equivalence. The underlying system Σ can be interpreted as the I-O system $\Sigma = (\mathbf{T}, W \oplus \{0\}, \mathcal{B})$ in which $\boldsymbol{w}$ is the output variable (Definition 3.27; the transfer function equals the zero mapping). An intuitive explanation of the fact that the pencil form provides a more natural description is that $\boldsymbol{w}$ "acts as an output" in the pencil form, whereas it "acts as an input" in the dual pencil form.

In case that there is an a priori decomposition $W = Y \oplus U$, it is always possible to obtain a D representation (DZ representation) from an AR or MA representation. The procedure is actually fairly simple: apply the above method to obtain a realization in pencil or dual pencil form and then apply either Algorithm 3.30 (Algorithm 3.32) or the dual version of this algorithm. The resulting D representation is, however, not necessarily minimal. In Subsection 5.3.2 we will give a method that yields a D representation (DZ representation) that is minimal under strong external equivalence. Since the concept of a transfer function has no specific meaning in the behavioral approach, we do not require the matrix $sE - A$ to be invertible. In fact, we allow the matrices E and A to be nonsquare.

For systems with nonproper transfer functions, one can obtain a realization in DZ form by the following trick, which appears at several places in the literature; see [12, 56, 70]. First finite and infinite frequencies are split by decomposing $T(s)$ as

$$T(s) = T_1(s) + T_2(s),$$

where $T_1(s)$ is strictly proper and $T_2(s)$ is polynomial. The second step consists of the realization of both $T_1(s)$ and $s^{-1}T_2(s^{-1})$ in standard state space form as

$$T_1(s) = C_1(sI - A_1)^{-1}B_1 \text{ and } s^{-1}T_2(s^{-1}) = C_2(sI - A_2)^{-1}B_2. \tag{5.1}$$

A small calculation shows that we then have $T_2(s) = C_2(I - sA_2)^{-1}B_2$ so that

$$T(s) = C(sE - A)^{-1}B,$$

where

$$E = \begin{bmatrix} I & 0 \\ 0 & -A_2 \end{bmatrix}, A = \begin{bmatrix} A_1 & 0 \\ 0 & I \end{bmatrix}, B = \begin{bmatrix} B_1 \\ B_2 \end{bmatrix}, C = [C_1 \quad C_2].$$

When the realizations in (5.1) are both minimal under transfer equivalence, then the *rank* of E is minimal among the DZ representations that realize $T(s)$. This has been proved in [56] and is essentially due to the fact that we then have rank $E = \delta^p(T)$; note that this is in accordance with (3.31). It is not difficult to prove that (E, A, B, C) is also minimal with respect to the *size* of E: the conditions of Theorem 4.29 are easily checked. Note, however, that the size of E is not necessarily minimal among the descriptor representations with *nonzero* direct feedthrough matrix, i.e., the D representations. In fact, we believe that a D realization for which the size of E is minimal cannot be obtained directly by separating finite and infinite frequencies. In contrast, the method that we will present in Subsection 5.3.2 yields a D representation (E, A, B, C, D) that is minimal with respect to the size of E.

In Section 5.4 we compare our method with the realization method of Fuhrmann [16, 18] that is also defined in a basis-free way. For this, we have to restrict ourselves to I-O systems with $W = Y \oplus U$ for which the transfer function $T(s)$ is strictly proper; the Fuhrmann realization method gives a standard state space realization $T(s) = C(sI - A)^{-1}B$ that is defined from a left polynomial factorization of $T(s)$. An important feature of the realization is that it is not controllable and therefore not minimal under transfer equivalence if the factorization that one starts with is not coprime. We will show that the Fuhrmann realization represents the *behavior* of the system that corresponds to the factorization and is minimal under *strong* external equivalence.

In the following section we present our basis-free realization method; the method yields a representation in pencil form that is minimal under strong external equivalence. The approach is based on results in [66] concerning the realization of a discrete-time behavior. Most of these results are spelled out in Section 5.2. In Section 5.3 we give a procedure for computing the pencil realization, which we illustrate by an example. In the same section we consider other variations on the theme: we will show how the computational procedure can be modified to obtain DP, D and DZ representations. In Subsection 5.3.1 realizations from MA representations are obtained by duality. In all algorithms the heaviest computational load consists of the inversion of a single constant $q \times q$-matrix. Some of the results of this chapter have been written down in [29, 30, 33].

5.1 Realization in pencil form: the abstract procedure

Let us consider a system $\Sigma = (\mathbf{T}, W, \mathcal{B}) \in \mathcal{L}^q$ whose behavior $\mathcal{B}$ is given by an AR representation

$$R(\sigma)w = 0, \tag{5.2}$$

where $R(s)$ is a polynomial matrix of size $r \times q$ of full row rank. Let $\mathcal{W}_0$ and W^0 be the RB space and the DV space of Σ, respectively, and let $\mathcal{W}_*$ be defined as in Definition 3.10; then $\mathcal{W}_0$ and $\mathcal{W}_*$ are given as

$$\mathcal{W}_0 = \{w(\lambda) \in \lambda^{-1}W_\infty(\lambda) \mid R(\lambda)w(\lambda) \text{ is polynomial } \}$$

$$\mathcal{W}_* = \{w(\lambda) \in \lambda^{-1}W_\infty(\lambda) \mid R(\lambda)w(\lambda) = 0\}.$$

We introduce the following space of polynomial vectors:

$$X_-(R) = \{p(\lambda) \in \mathbb{R}^r[\lambda] \mid \exists w(\lambda) \in \mathcal{W}_0 \text{ such that } p(\lambda) = R(\lambda)w(\lambda)\}.$$

The space $X_-(R)$ is finite-dimensional; it follows from Remark 2.20 that

$$\dim X_-(R) = \kappa^\ell(R) = \text{ord}(\Sigma). \tag{5.3}$$

The space $X_-(R)$ is clearly isomorphic to the quotient space $\mathcal{W}_0/\mathcal{W}_*$; denoting an element of the quotient space by $[\ .\ ]$ we find a natural isomorphism in the mapping M_R from $\mathcal{W}_0/\mathcal{W}_*$ to $X_-(R)$ that is defined by

$$M_R : [w(\lambda)] \mapsto R(\lambda)w(\lambda).$$

The quotient space $\mathcal{W}_0/\lambda^{-1}\mathcal{W}_*$ is also finite-dimensional; it is easily checked that

$$\begin{aligned} \dim \mathcal{W}_0/\lambda^{-1}\mathcal{W}_* &= \dim \mathcal{W}_0/\mathcal{W}_* + \dim W^0 \\ &= \text{ord}(\Sigma) + \dim W^0. \end{aligned} \tag{5.4}$$

Next, we define mappings G and F from $\mathcal{W}_0/\lambda^{-1}\mathcal{W}_*$ to $X_-(R)$ by

$$G : [w(\lambda)] \mapsto R(\lambda)w(\lambda) \tag{5.5}$$

$$F : [w(\lambda)] \mapsto R(\lambda)\pi_-(\lambda w(\lambda)) \tag{5.6}$$

and a mapping H from $\mathcal{W}_0/\lambda^{-1}\mathcal{W}_*$ to W by

$$H : [w(\lambda)] \mapsto \pi_1 w(\lambda). \tag{5.7}$$

Here π_- denotes the projection of $W(\lambda)$ onto $\lambda^{-1}W_\infty(\lambda)$, effected by "deleting the polynomial part", and π_1 is defined as in (2.23).

At this moment it is already clear that (F, G, H) is minimal under strong external equivalence. Indeed, it follows from (5.3) and (5.4) that the lower bounds of Theorem 4.3 are reached. Hence we essentially have to prove only one thing: (F, G, H) and (5.2) represent the same behavior (note the difference in strategy with [29]).

Theorem 5.1 Let $R(s)$ be a polynomial matrix of size $r \times q$ that has full row rank. Let F, G and H be defined by (5.5–5.7). Then the P representation (F, G, H) is strongly externally equivalent to the AR representation

$$R(\sigma)w = 0. \tag{5.8}$$

Moreover, the representation (F, G, H) is minimal under strong external equivalence, whereas it is minimal under weak external equivalence if and only if $R(s)$ has full row rank for all $s \in \mathbb{C}$.

Proof We first prove that the representations are strongly externally equivalent. According to Theorem 3.12 it suffices to show that the RB space corresponding to (F, G, H) and the RB space corresponding to (5.8) are the same. For this, let $w(s)$ be an element of the RB space corresponding to (5.8), i.e., $w(s) \in s^{-1}W_\infty(s)$ and $R(s)w(s)$ is polynomial. We will show that $w(s)$ belongs to the RB space corresponding to (F, G, H). By Lemma 3.16 it suffices to find an element $\xi(s) \in s^{-1}Z_\infty(s)$ such that $(sG-F)\xi(s)$ is polynomial and $w(s) = H\xi(s)$. Define $\xi_0 \in Z$ by $\xi_0 := [w(\lambda)]$ (here $[\,.\,]$ denotes the equivalence class in $\mathcal{W}_0/\lambda^{-1}\mathcal{W}_*$). We now define $\xi(s)$ by

$$\xi(s) := (sI - P)^{-1}\xi_0,$$

where P is the mapping from Z to Z that is defined by

$$P : [w(\lambda)] \mapsto [\pi_-\lambda w(\lambda)].$$

Clearly $\xi(s) \in s^{-1}Z_\infty(s)$ and $w(s) = H\xi(s)$. From the definition of F and G it follows that $(sG - F)\xi(s) = G\xi_0$ is a constant, so $\xi(s)$ indeed has the desired properties. Conversely, let $w(s)$ be an element of the RB space corresponding to (F, G, H) so that there exists an element $\xi(s) \in Z(s)$ such that $(sG - F)\xi(s)$ is polynomial and $w(s) = \pi_-(H\xi(s))$. Without loss of generality, we may assume that $\xi(s)$ is strictly proper so that $w(s) = H\xi(s)$ and $(sG - F)\xi(s)$ is a constant, say $p(\lambda) \in X_-(R)$. We will prove that $R(s)w(s)$ is polynomial; more specifically we will prove that $R(s)w(s) = p(s)$. To indicate the dependance on λ, we will denote $\xi(s)$ as $\xi(\lambda)(s)$ in the sequel. Let us write the Laurent expansion of $\xi(\lambda)(s)$ as

$$\xi(\lambda)(s) = [w_1(\lambda)]s^{-1} + [w_2(\lambda)]s^{-2} + \cdots.$$

From $(sG - F)\xi(\lambda)(s) = p(\lambda)$ it now follows that

$$G([w_1(\lambda)]) = R(\lambda)w_1(\lambda) = p(\lambda) \tag{5.9}$$

and that

$$G([w_{i+1}(\lambda)]) = F([w_i(\lambda)]) \text{ for all } i \in \mathbb{Z}_+,$$

which implies that

$$R(\lambda)w_{i+1}(\lambda) = \lambda R(\lambda)w_i(\lambda) - R(\lambda)(H[w_i(\lambda)]) \text{ for all } i \in \mathbb{Z}_+. \tag{5.10}$$

The Laurent expansion of the vector $R(s)w(s)$ can be written as

$$\begin{aligned} R(s)w(s) &= R(s)(H[w_1(\lambda)])s^{-1} + R(s)(H[w_2(\lambda)])s^{-2} + \cdots \\ &= R(s)(H[w_1(s)])s^{-1} + R(s)(H[w_2(s)])s^{-2} + \cdots. \end{aligned}$$

Substituting $\lambda = s$ in equation (5.10) we then get

$$R(s)w(s) = R(s)w_1(s),$$

which equals $p(s)$ by (5.9). This proves that $w(s)$ belongs to the RB space corresponding to (5.8) and we conclude that the two representations are strongly externally equivalent. The statement on minimality under strong external equivalence is already implied by the remarks preceding the theorem. To prove the statement on minimality under weak external equivalence, we first note that, under the assumption that $R(s)$ has full row rank, it follows from the proof of Lemma 2.28 that $\kappa^{\ell}(R) = \nu(\ker R(s))$ if and only if $R(s)$ has full row rank for all $s \in \mathbb{C}$. Since, by definition, $\nu(\ker R(s))$ equals C-ord(Σ), the statement follows from the above and Theorem 4.7.

5.2 The pencil realization in terms of a discrete-time behavior

In the previous section we presented a method for realization in pencil form of systems that were given by an AR representation; the pencil matrices F, G and H are computed from the polynomial matrix in the AR representation, and the method is valid in continuous as well as in discrete time. In this section we will show that for the discrete-time case the realization can also be formulated in terms of the behavior itself. This will shed some light on the use of certain spaces in the previous sections; the link is provided by the correspondence

$$(w_0, w_1, \ldots) \leftrightarrow w_0\lambda^{-1} + w_1\lambda^{-2} + \cdots,$$

which identifies $W^{\mathbb{Z}_+}$ with the set of formal power series in λ with vanishing constant term. Our treatment here is based on the development in [66]; however, we derive some results, published in [29], that do not depend on the assumption that the behavior is closed in the topology of pointwise convergence.

We already introduced the mapping σ on $W^{\mathbb{Z}_+}$ in Section 3.1; it was defined as the (backward) shift

$$\sigma : (w_0, w_1, \ldots) \mapsto (w_1, w_2, \ldots).$$

We shall also consider the following mappings on $W^{\mathbb{Z}_+}$: the *forward shift*

$$\sigma^* : (w_0, w_1, \ldots) \mapsto (0, w_0, w_1, \ldots),$$

and the *evaluation mapping at time 0*

$$\chi : (w_0, w_1, \ldots) \mapsto w_0.$$

Let us consider a subspace $\mathcal{B}$ of $W^{\mathbb{Z}_+}$ that is linear and σ-invariant. We introduce, for each $k \in \mathbb{Z}_+$, the following spaces:

$$\mathcal{B}_k = \{[w]_k \mid w \in \mathcal{B}\},$$

where $[w]_k$ denotes the *k-truncation* of an element w of $W^{\mathbb{Z}_+}$: if

$$w = (w_0, w_1, \ldots, w_k, w_{k+1}, \ldots),$$

then

$$[w]_k = (w_0, w_1, \ldots, w_k).$$

Recall that pointwise convergence on $W^{\mathbb{Z}_+}$ is defined as

$$w^i \overset{i\to\infty}{\longrightarrow} w \iff \|w^i_k - w_k\| \overset{i\to\infty}{\longrightarrow} 0 \text{ for all } k \in \mathbb{Z}_+,$$

where $\|.\|$ denotes the usual Euclidean norm on W. Denoting the closure of $\mathcal{B}$ in the topology of pointwise convergence by $\mathcal{B}^{cl}$, we have the following lemma:

Lemma 5.2 Let $\mathcal{B} \subset W^{\mathbb{Z}_+}$ be linear and σ-invariant. Then

$$w \in \mathcal{B}^{cl} \iff [w]_k \in \mathcal{B}_k \text{ for all } k \in \mathbb{Z}_+.$$

Proof See the proof of Proposition 4 in [66]. ◇

Following [66] we introduce the subspaces

$$\mathcal{B}^0 = \{w \in \mathcal{B} \mid (\sigma^*)^k w \in \mathcal{B} \text{ for all } k \in \mathbb{Z}_+\} \tag{5.11}$$

and

$$\mathcal{B}^1 = \{w \in \mathcal{B}^0 \mid \chi w = 0\}. \tag{5.12}$$

Note that $\mathcal{B}^0$ is, by definition, the largest σ^*-invariant subspace of $\mathcal{B}$. Intuitively, $\mathcal{B}^0$ contains the trajectories that start from zero "state"; so the quotient space $\mathcal{B}/\mathcal{B}^0$ can be interpreted as a state space. The quotient space $\mathcal{B}^0/\mathcal{B}^1$ describes the freedom one has in choosing a value of the "driving variable" of the system. As a result, $\mathcal{B}^0/\mathcal{B}^1$ can be interpreted as the space of "driving variables". The natural candidate for the internal variable space Z is therefore the quotient space $\mathcal{B}/\mathcal{B}^1$, which is isomorphic to $\mathcal{B}/\mathcal{B}^0 \oplus \mathcal{B}^0/\mathcal{B}^1$. The following facts are trivially verified:

$$\sigma\mathcal{B}^1 \subset \mathcal{B}^0 \tag{5.13}$$

$$\mathcal{B}^1 \subset \ker \chi. \tag{5.14}$$

Because of (5.13), we can properly define a mapping $M_1 : \mathcal{B}/\mathcal{B}^1 \to \mathcal{B}/\mathcal{B}^0$ by

$$M_1 : w \bmod \mathcal{B}^1 \mapsto \sigma w \bmod \mathcal{B}^0. \tag{5.15}$$

Because of (5.14), there is also a mapping $M_2 : \mathcal{B}/\mathcal{B}^1 \to W$ defined by

$$M_2 : w \bmod \mathcal{B}^1 \mapsto \chi w. \tag{5.16}$$

Finally, because $\mathcal{B}^1 \subset \mathcal{B}^0$, we can introduce the projection mapping $M_0 : \mathcal{B}/\mathcal{B}^1 \to \mathcal{B}/\mathcal{B}^0$, defined simply by

$$M_0 : \; \boldsymbol{w} \bmod \mathcal{B}^1 \mapsto \boldsymbol{w} \bmod \mathcal{B}^0. \tag{5.17}$$

The mappings M_0, M_1 and M_2 could also have been introduced by requiring that the diagram in Figure 5.1 below commutes; in the diagram π^0 denotes projection modulo $\mathcal{B}^0$, whereas π^1 denotes projection modulo $\mathcal{B}^1$. In the sequel the discrete-time behavior that corresponds to a pencil

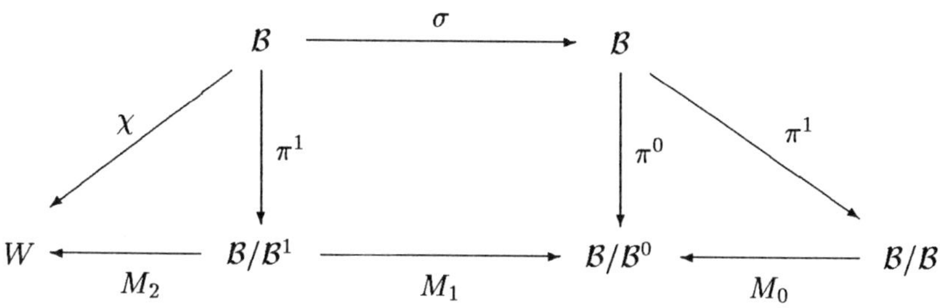

Figure 5.1: Pencil realization defined directly from the behavior

representation (F, G, H), where F and G are mappings from Z to X and H is a mapping from Z to W is denoted by $\mathcal{B}_p(Z, X, W; F, G, H)$; following Definition 3.13 we have

$$\mathcal{B}_p(Z, X, W; F, G, H) = \{\boldsymbol{w} : \mathbb{Z}_+ \to W \mid \exists \boldsymbol{z} : \mathbb{Z}_+ \to Z \text{ such that}$$
$$\sigma G \boldsymbol{z} = F\boldsymbol{z} \text{ and } \boldsymbol{w} = H\boldsymbol{z}\}.$$

We can now formulate the following theorem:

Theorem 5.3 Let $\mathcal{B} \subset W^{\mathbb{Z}_+}$ be linear and σ-invariant. Let the spaces $\mathcal{B}^0$ and $\mathcal{B}^1$ and the mappings M_0, M_1, and M_2 be defined by (5.11), (5.12), (5.17), (5.15), and (5.16), respectively. Then

$$\mathcal{B} \subset \mathcal{B}_p(\mathcal{B}/\mathcal{B}^1, \mathcal{B}/\mathcal{B}^0, W; M_1, M_0, M_2) \subset \mathcal{B}^{cl}.$$

Proof In order to prove that $\mathcal{B} \subset \mathcal{B}_p(\mathcal{B}/\mathcal{B}^1, \mathcal{B}/\mathcal{B}^0, W; M_1, M_0, M_2)$, let us assume that $\boldsymbol{w} \in \mathcal{B}$. Define $\boldsymbol{z} : \mathbf{Z}_+ \to \mathcal{B}/\mathcal{B}^1$ by

$$z_k = \pi^1 \sigma^k \boldsymbol{w}.$$

One easily verifies from the definitions that $\sigma M_0 \boldsymbol{z} = M_1 \boldsymbol{z}$ and that $M_2 \boldsymbol{z} = \boldsymbol{w}$. This proves that $\boldsymbol{w} \in \mathcal{B}_p(\mathcal{B}/\mathcal{B}^1, \mathcal{B}/\mathcal{B}^0, W; M_1, M_0, M_2)$.

To show that $\mathcal{B}_p(\mathcal{B}/\mathcal{B}^1, \mathcal{B}/\mathcal{B}^0, W; M_1, M_0, M_2) \subset \mathcal{B}^{cl}$, let us assume that $\boldsymbol{w} \in \mathcal{B}_p(\mathcal{B}/\mathcal{B}^1, \mathcal{B}/\mathcal{B}^0, W; M_1, M_0, M_2)$. Because of Lemma 5.2, it suffices to prove that $[\boldsymbol{w}]_k \in \mathcal{B}_k$ for all $k \in \mathbb{Z}_+$. Let $\boldsymbol{z} : \mathbf{Z}_+ \to \mathcal{B}/\mathcal{B}^1$ be such that

$\sigma M_0 \boldsymbol{z} = M_1 \boldsymbol{z}$ and $\boldsymbol{w} = M_2 \boldsymbol{z}$. Since $\pi^1 : \mathcal{B} \to \mathcal{B}/\mathcal{B}^1$ is surjective, we can find for every k a $\hat{\boldsymbol{w}}^k \in \mathcal{B}$ such that

$$z_k = \pi^1 \hat{\boldsymbol{w}}^k.$$

Define $\tilde{\boldsymbol{w}}^k$ by

$$\tilde{\boldsymbol{w}}^k := (\hat{w}_0^0, \hat{w}_0^1, \ldots, \hat{w}_0^k, \hat{w}_1^k, \hat{w}_2^k, \ldots).$$

Then $[\tilde{\boldsymbol{w}}]_k = [\boldsymbol{w}]_k$ for all $k \in \mathbb{Z}_+$ because of the fact that

$$w_k = M_2 z_k = M_2 \pi^1 \hat{\boldsymbol{w}}^k = \chi \hat{\boldsymbol{w}}^k = \hat{w}_0^k \text{ for all } k \in \mathbb{Z}_+.$$

It remains to prove that $\tilde{\boldsymbol{w}}^k \in \mathcal{B}$ for all $k \in \mathbb{Z}_+$. For $k = 0$, this is trivial since $\tilde{\boldsymbol{w}}^0 = \hat{\boldsymbol{w}}^0 \in \mathcal{B}$. Because

$$\begin{aligned} \tilde{\boldsymbol{w}}^{k+1} - \tilde{\boldsymbol{w}}^k &= (0, 0, \ldots, 0, \hat{w}_0^{k+1} - \hat{w}_1^k, \hat{w}_1^{k+1} - \hat{w}_2^k, \ldots) = \\ &= (\sigma^*)^k (\hat{\boldsymbol{w}}^{k+1} - \sigma \hat{\boldsymbol{w}}^k), \end{aligned} \tag{5.18}$$

the proof will follow by induction if we can show that

$$\hat{\boldsymbol{w}}^{k+1} - \sigma \hat{\boldsymbol{w}}^k \in \mathcal{B}^0 \text{ for all } k \in \mathbb{Z}_+.$$

But this follows from

$$\pi^0 \hat{\boldsymbol{w}}^{k+1} = M_0 \pi^1 \hat{\boldsymbol{w}}^{k+1} = M_0 z_{k+1} = M_1 z_k = M_1 \pi^1 \hat{\boldsymbol{w}}^k = \pi^0 \sigma \hat{\boldsymbol{w}}^k.$$

◇

The above theorem leads to the following result, which coincides with Theorem 9 in [66].

Corollary 5.4 Let $\mathcal{B} \subset W^{\mathbb{Z}_+}$ be linear, σ-invariant and closed. Then

$$\mathcal{B}_p(\mathcal{B}/\mathcal{B}^1, \mathcal{B}/\mathcal{B}^0, W; M_1, M_0, M_2) = \mathcal{B}.$$

◇

It was shown in [66] that the spaces $\mathcal{B}/\mathcal{B}^0$ and $\mathcal{B}/\mathcal{B}^1$ are finite-dimensional if $\mathcal{B}$ is linear, σ-invariant and closed. For completeness, we shall offer a proof of this fact which we think is more straightforward. We first define a sequence of subspaces of W by

$$W_k^0(\mathcal{B}) := \{w \in W \mid (0, 0, \ldots, 0, w) \in \mathcal{B}_k\}.$$

It is immediate from $\sigma\mathcal{B} \subset \mathcal{B}$ that $W_{k+1}^0(\mathcal{B}) \subset W_k^0(\mathcal{B})$ for all k. Because W is finite-dimensional, the sequence of subspaces $W_0^0(\mathcal{B}) \supset W_1^0(\mathcal{B}) \supset \cdots$ must reach a limit after a finite number of steps; the limit subspace will be denoted by $W^0(\mathcal{B})$. We have the following lemma:

Lemma 5.5 Let $\mathcal{B} \subset W^{\mathbb{Z}_+}$ be linear, σ-invariant and closed. Let k_0 be such that $W_{k_0}^0(\mathcal{B}) = W^0(\mathcal{B})$, and let $\Phi : \mathcal{B} \to \mathcal{B}_{k_0}$ denote the mapping $\boldsymbol{w} \mapsto [\boldsymbol{w}]_{k_0}$. Then

$$\ker \Phi \subset \mathcal{B}^0.$$

Proof Since $\mathcal{B}^0$ is by definition the largest σ^*-invariant subspace of $\mathcal{B}$, it suffices to show that $\ker \Phi$ is σ^*-invariant. Take $\boldsymbol{w} \in \ker \Phi$; we want to show that also $\sigma^* \boldsymbol{w} \in \ker \Phi$, which will follow if we can prove that $\sigma^* \boldsymbol{w} \in \mathcal{B}$. By the closedness of $\mathcal{B}$, it is sufficient to show (Lemma 5.2) that

$$[\sigma^* \boldsymbol{w}]_k \in \mathcal{B}_k \text{ for all } k \in \mathbb{Z}_+.$$

For $0 \leq k \leq k_0 + 1$, $[\sigma^* \boldsymbol{w}]_k = 0$, and so the above condition is certainly satisfied. To proceed by induction, suppose that $[\sigma^* \boldsymbol{w}]_i \in \mathcal{B}_i$ for some $i \geq k_0 + 1$. Let $\tilde{\boldsymbol{w}} \in \mathcal{B}$ be such that $[\tilde{\boldsymbol{w}}]_i = [\sigma^* \boldsymbol{w}]_i$. We then have $[\boldsymbol{w} - \sigma \tilde{\boldsymbol{w}}]_{i-1} = 0$, and therefore

$$w_i - \tilde{w}_{i+1} \in W_i^0(\mathcal{B}) = W_{i+1}^0(\mathcal{B}). \tag{5.19}$$

From (5.19) and the fact that $[\sigma^* \boldsymbol{w} - \tilde{\boldsymbol{w}}]_i = 0$, it follows that

$$[\sigma^* \boldsymbol{w} - \tilde{\boldsymbol{w}}]_{i+1} \in \mathcal{B}_{i+1}.$$

Since $[\tilde{\boldsymbol{w}}]_{i+1}$ obviously belongs to $\mathcal{B}_{i+1}$, we can conclude that $[\sigma^* \boldsymbol{w}]_{i+1} \in \mathcal{B}_{i+1}$, which is what we wanted to prove.

Remark 5.6 From the lemma one easily derives that $W^0(\mathcal{B})$ coincides with the space $\chi \mathcal{B}^0$.

Theorem 5.7 Let $\mathcal{B} \subset W^{\mathbb{Z}_+}$ be linear, σ-invariant and closed. Then the spaces $\mathcal{B}/\mathcal{B}^0$ and $\mathcal{B}/\mathcal{B}^1$ are finite-dimensional.

Proof By the previous lemma, we have

$$\dim \mathcal{B}/\mathcal{B}^0 \leq \dim \mathcal{B}/\ker \Phi = \dim \text{ im } \Phi \leq \dim W^{k_0+1} = q(k_0+1).$$

Because $\dim \mathcal{B}/\mathcal{B}^1 = \dim \mathcal{B}/\mathcal{B}^0 + \dim \mathcal{B}^0/\mathcal{B}^1$, this also implies that $\mathcal{B}/\mathcal{B}^1$ is finite-dimensional, since it follows immediately from the definitions that $\dim \mathcal{B}^0/\mathcal{B}^1 \leq \dim W$. ◇

Below we give a direct proof of the fact that the pencil representation obtained above is minimal; this is an immediate result of the fact that $\mathcal{B}^0$ is the *largest* σ^*-invariant subspace of $\mathcal{B}$.

Theorem 5.8 If $(Z, X, W; F, G, H)$ is a pencil representation of the linear, time-invariant behavior $\mathcal{B}$, then

$$\dim X \geq \dim \mathcal{B}/\mathcal{B}^0$$

and

$$\dim Z \geq \dim \mathcal{B}/\mathcal{B}^1.$$

Proof Define the space $\mathcal{B}_z \subset Z^{\mathbb{Z}_+}$ by

$$\mathcal{B}_z := \{\boldsymbol{z} : \mathbb{Z}_+ \to Z \mid \sigma G \boldsymbol{z} = F \boldsymbol{z}\}.$$

By definition of a pencil representation, we have

$$H\mathcal{B}_z = \mathcal{B}. \tag{5.20}$$

In analogy with $\mathcal{B}^0$, we also introduce

$$\mathcal{B}_z^0 = \{\boldsymbol{z} \in \mathcal{B}_z \mid (\sigma^*)^k \boldsymbol{z} \in \mathcal{B}_z \quad \text{for all } k \geq 0\}.$$

Obviously, one has

$$H\mathcal{B}_z^0 \subset \mathcal{B}^0. \tag{5.21}$$

It is easily verified that, in fact,

$$\mathcal{B}_z^0 = \{\boldsymbol{z} \in \mathcal{B}_z \mid G z_0 = 0\},$$

which shows that $\mathcal{B}_z^0$ is the kernel of the mapping which assigns the element Gz_0 of X to a given $\boldsymbol{z} \in \mathcal{B}_z$. As a consequence, we get

$$\dim(\mathcal{B}_z/\mathcal{B}_z^0) \leq \dim X.$$

Because of (5.21), we can unambiguously define a mapping $\Psi : \mathcal{B}_z/\mathcal{B}_z^0 \to \mathcal{B}/\mathcal{B}^0$ by

$$\Psi : \boldsymbol{z} \bmod \mathcal{B}_z^0 \mapsto H\boldsymbol{z} \bmod \mathcal{B}^0.$$

Moreover, (5.20) shows that this map is surjective. Therefore,

$$\dim \mathcal{B}/\mathcal{B}^0 \leq \dim \mathcal{B}_z/\mathcal{B}_z^0 \leq \dim X.$$

For the proof of the second inequality, one introduces

$$\mathcal{B}_z^1 = \{\boldsymbol{z} \in \mathcal{B}_z^0 \mid z_0 = 0\} = \{\boldsymbol{z} \in \mathcal{B}_z \mid z_0 = 0\}$$

and proceeds analogously, noting that $H\mathcal{B}_z^1 \subset \mathcal{B}^1$ and that

$$\dim(\mathcal{B}_z/\mathcal{B}_z^1) \leq \dim Z.$$

5.3 Choosing bases

In this section we shall construct specific bases for the spaces that appeared in the abstract realization of Section 5.1. In this way we obtain concrete matrices F, G and H corresponding to a pencil realization. The procedure will be illustrated by an example. In subsequent subsections we show how the computational procedure can be modified to obtain DP, D and DZ representations. In Subsection 5.3.1 realizations from MA representations are obtained by duality.

We now present the abovementioned procedure to obtain pencil matrices F, G and H from an AR representation given by a row reduced $r \times q$ polynomial matrix $R(s)$.

Procedure 5.9 The first step is to factorize the matrix $R(s)$ as in Definition 2.18 as

$$R(s) = \Delta(s)B(s) \text{ with } \Delta(s) = \operatorname{diag}(s^{\kappa_1}, \ldots, s^{\kappa_r}), \tag{5.22}$$

where $B(s)$ is right $\mathbb{R}_\infty(s)$-unimodular and the κ_i's are the row degrees of $R(s)$; by Lemma 2.22 they coincide with the observability indices of Σ. Note that we have $\kappa_1 + \kappa_2 + \cdots + \kappa_r = \operatorname{ord}(\Sigma)$ and $q - r = \dim W^0$.

The second step consists of choosing a matrix $\tilde{B}(s)$ such that

$$\hat{B}(s) := \begin{bmatrix} B(s) \\ \tilde{B}(s) \end{bmatrix}$$

is $\mathbb{R}_\infty(s)$-unimodular. We may then write $R(s) = [\Delta(s) \quad 0]\,\hat{B}(s)$, and it is seen from this that a basis for $\mathcal{W}_0/\lambda^{-1}\mathcal{W}_*$ is given by the equivalence classes modulo $\lambda^{-1}\mathcal{W}_*$ of the columns of the following matrix of size $q \times (\kappa_1 + \kappa_2 + \cdots + \kappa_r + (q - r))$:

$$\hat{B}^{-1}(\lambda)\begin{bmatrix} \lambda^{-1} & \cdots & \lambda^{-\kappa_1} & & & & & & \\ & & & \cdots & & & & & \\ & & & & \lambda^{-1} & \cdots & \lambda^{-\kappa_r} & & \\ & & & & & & & \lambda^{-1} & & \\ & & & & & & & & \ddots & \\ & & & & & & & & & \lambda^{-1} \end{bmatrix}.$$

A basis matrix for $X_-(R)$ is given by the following matrix of size $r \times (\kappa_1 + \kappa_2 + \cdots + \kappa_r)$:

$$\begin{bmatrix} \lambda^{\kappa_1 - 1} & \cdots & \lambda & 1 & & & & & \\ & & & & \cdots & & & & \\ & & & & & \cdots & & & \\ & & & & & & \lambda^{\kappa_r - 1} & \cdots & \lambda & 1 \end{bmatrix}.$$

With respect to these bases, we now compute the matrix forms of the mappings F, G and H as defined by (5.5–5.7). It is easily seen that G will take the form

$$G = [I \quad 0],$$

where the partitioning is $\operatorname{ord}(\Sigma) \times (\operatorname{ord}(\Sigma) + (q - r))$. Because $\hat{B}(\lambda)$ is $\mathbb{R}_\infty(\lambda)$-unimodular, the matrix of H will have the form

$$H = \hat{B}(\infty)^{-1}\begin{bmatrix} 1 & \cdots & 0 & & & & & & \\ & & & \cdots & & & & & \\ & & & & 1 & \cdots & 0 & & \\ & & & & & & & 1 & & \\ & & & & & & & & \ddots & \\ & & & & & & & & & 1 \end{bmatrix},$$

where the partitioning is $q \times (\kappa_1 + \kappa_2 + \cdots + \kappa_r + (q-r))$. Here, we see that we need the inverse of $\hat{B}(\infty)$. Finally, if we let $G(\lambda)$ denote the matrix whose columns are the images under G in $\mathbb{R}^r[\lambda]$ of the basis elements for $\mathcal{W}_0/\lambda^{-1}\mathcal{W}_*$ displayed above, then we can compute a similar matrix for F by the formula

$$F(\lambda) = \lambda G(\lambda) - R(\lambda)H,$$

which follows from the definitions of F, G, and H. This is easily transformed to a matrix expression for F because of the simple basis we chose for $X_-(R)$. ◇

The following theorem is an immediate result of Theorem 5.1.

Theorem 5.10 Let $R(s)$ be a polynomial matrix of size $r \times q$ that is row reduced. Let F, G and H be defined as in Procedure 5.9. Then the P representation (F, G, H) is strongly externally equivalent to the AR representation

$$R(\sigma)w = 0.$$

Moreover, the representation (F, G, H) is minimal under strong external equivalence, whereas it is minimal under weak external equivalence if and only if $R(s)$ has full row rank for all $s \in \mathbb{C}$.

Remark 5.11 When the polynomial matrix $R(s)$ is factorized as in (5.22), it follows from the definition that the DV space W^0 of Σ is given as

$$W^0 = \ker B(\infty).$$

Of course, this can also be written as

$$W^0 = \operatorname{im} \hat{B}(\infty)^{-1} \begin{bmatrix} 0 \\ I_{(q-r)\times(q-r)} \end{bmatrix}$$

so that it is immediate from the definition of G and H that $W^0 = H[\ker G]$, which is in accordance with Lemma 4.5.

Example 5.12 The mass-spring-damper system of Example 1.3 is given by the AR representation

$$R(\frac{d}{dt}) \begin{bmatrix} q \\ F \end{bmatrix} = 0$$

with $R(s) = \begin{bmatrix} Ms^2 + cs + k & -1 \end{bmatrix}$. Since $R(s)$ has only one row, we can immediately write down a minimal DP representation by using the simple realization method given at the beginning of this chapter. This leads to the DP representation

$$\begin{bmatrix} 1 & 0 \\ 0 & 1 \\ 0 & 0 \end{bmatrix} \begin{bmatrix} \dot{\xi}_1 \\ \dot{\xi}_2 \end{bmatrix} = \begin{bmatrix} 0 & 0 \\ 1 & 0 \\ 0 & 1 \end{bmatrix} \begin{bmatrix} \xi_1 \\ \xi_2 \end{bmatrix} + \begin{bmatrix} k & -1 \\ c & 0 \\ M & 0 \end{bmatrix} \begin{bmatrix} \boldsymbol{q} \\ \boldsymbol{F} \end{bmatrix}.$$

To illustrate our realization method, we will now realize this system in pencil form as well. The matrix $R(s)$ is row reduced with row degree 2: $R(s) = s^2 B(s)$ with $B(\infty) = [M \quad 0]$. A polynomial basis matrix for $X_-(R)$ is given by $[\lambda \quad 1]$. Furthermore, dim ker $R(s) = q - r = 2 - 1 = 1$, so we get $G = [I \quad 0] \in \mathbb{R}^{2\times 3}$. We now have to choose $\tilde{B}$ to complete $B(\infty)$ to a nonsingular matrix; we choose $\tilde{B} = [0 \quad 1]$ so that

$$\hat{B}(\infty) = \begin{bmatrix} M & 0 \\ 0 & 1 \end{bmatrix}, \quad \hat{B}(\infty)^{-1} = \frac{1}{M} \begin{bmatrix} 1 & 0 \\ 0 & M \end{bmatrix}.$$

Therefore,

$$H = \frac{1}{M} \begin{bmatrix} 1 & 0 \\ 0 & M \end{bmatrix} \begin{bmatrix} 1 & 0 & 0 \\ 0 & 0 & 1 \end{bmatrix} = \frac{1}{M} \begin{bmatrix} 1 & 0 & 0 \\ 0 & 0 & M \end{bmatrix}.$$

Finally,

$$F(\lambda) = \begin{bmatrix} \lambda^2 & \lambda & 0 \end{bmatrix} - R(\lambda)H = \frac{1}{M} \begin{bmatrix} -c\lambda - k & M\lambda & M \end{bmatrix}.$$

The matrix of F is, therefore,

$$F = \frac{1}{M} \begin{bmatrix} -c & M & 0 \\ -k & 0 & M \end{bmatrix}.$$

The resulting minimal P representation is:

$$\begin{bmatrix} 1 & 0 & 0 \\ 0 & 1 & 0 \end{bmatrix} \begin{bmatrix} \dot{\xi}_1 \\ \dot{\xi}_2 \\ \dot{\xi}_3 \end{bmatrix} = \frac{1}{M} \begin{bmatrix} -c & M & 0 \\ -k & 0 & M \end{bmatrix} \begin{bmatrix} \xi_1 \\ \xi_2 \\ \xi_3 \end{bmatrix}$$

$$\begin{bmatrix} \boldsymbol{q} \\ \boldsymbol{F} \end{bmatrix} = \frac{1}{M} \begin{bmatrix} 1 & 0 & 0 \\ 0 & 0 & M \end{bmatrix} \begin{bmatrix} \xi_1 \\ \xi_2 \\ \xi_3 \end{bmatrix}.$$

Note that a different choice of bases for the spaces that appear in the abstract realization of Section 5.1 could, of course, lead to different matrices F, G and H. To illustrate this, let us take a basis for $X_-(R)$ different than the one in Procedure 5.1: let a polynomial basis matrix for $X_-(R)$ be $\begin{bmatrix} M\lambda + c & 1 \end{bmatrix}$. Let us then write

$$R(s) = \begin{bmatrix} Ms^2 + cs + k & 0 \end{bmatrix} B(s)$$

with

$$B(\infty) = \begin{bmatrix} 1 & 0 \\ 0 & 1 \end{bmatrix}.$$

Proceeding further as in Procedure 5.1, we derive the minimal P representation

$$\begin{bmatrix} 1 & 0 & 0 \\ 0 & 1 & 0 \end{bmatrix} \begin{bmatrix} \dot{\xi}_1 \\ \dot{\xi}_2 \\ \dot{\xi}_3 \end{bmatrix} = \frac{1}{M} \begin{bmatrix} 0 & 1 & 0 \\ -kM & -c & M \end{bmatrix} \begin{bmatrix} \xi_1 \\ \xi_2 \\ \xi_3 \end{bmatrix}$$

$$\begin{bmatrix} \boldsymbol{q} \\ \boldsymbol{F} \end{bmatrix} = \begin{bmatrix} 1 & 0 & 0 \\ 0 & 0 & 1 \end{bmatrix} \begin{bmatrix} \xi_1 \\ \xi_2 \\ \xi_3 \end{bmatrix}.$$

Example 5.13 Consider the two connected mass-spring-damper systems of Example 1.2 with $M_1 = M_2 = 1$, which are given by an AR representation with

$$R(s) = \begin{bmatrix} s^2 + c_1 s + k_1 & 0 & -1 & 0 \\ 0 & s^2 + c_2 s + k_2 & 0 & -1 \\ 0 & 0 & 1 & 1 \\ 1 & -1 & 0 & 0 \end{bmatrix}.$$

The matrix $R(s)$ is not row reduced, so we first have to find a unimodular polynomial matrix $U(s)$ for which $U(s)R(s)$ is row reduced. We use the procedure in the proof of Theorem 2.19 for determining $U(s)$. We then have to make a distinction into two cases (note that in practice this accounts for numerical difficulties):

CASE I: $c_1 = c_2$

The matrix

$$U(s) = \begin{bmatrix} 0 & 1 & 0 & 0 \\ 1 & -1 & 0 & -s^2 - c_1 s \\ 0 & 0 & 1 & 0 \\ 0 & 0 & 0 & 1 \end{bmatrix}$$

is unimodular and has the property that $\tilde{R}(s) = U(s)R(s)$ is row reduced with row degrees 2, 0, 0, and 0:

$$\tilde{R}(s) = \begin{bmatrix} 0 & s^2 + c_1 s + k_2 & 0 & -1 \\ k_1 & -k_2 & -1 & 1 \\ 0 & 0 & 1 & 1 \\ 1 & -1 & 0 & 0 \end{bmatrix} = \operatorname{diag}(s^2, 1, 1, 1) B(s),$$

where

$$B(\infty) = \begin{bmatrix} 0 & 1 & 0 & 0 \\ k_1 & -k_2 & -1 & 1 \\ 0 & 0 & 1 & 1 \\ 1 & -1 & 0 & 0 \end{bmatrix}$$

so that

$$B(\infty)^{-1} = \frac{1}{2} \begin{bmatrix} 2 & 0 & 0 & 2 \\ 2 & 0 & 0 & 0 \\ k_1 - k_2 & -1 & 1 & k_1 \\ -(k_1 - k_2) & 1 & 1 & -k_1 \end{bmatrix}.$$

We have dim ker $\tilde{R}(s) = q - r = 4 - 4 = 0$, so we get $G = I \in \mathbb{R}^{2\times 2}$. A polynomial basis matrix for $X_-(\tilde{R})$ is given by

$$\begin{bmatrix} \lambda & 1 \\ 0 & 0 \\ 0 & 0 \\ 0 & 0 \end{bmatrix}.$$

We now have

$$H = \frac{1}{2} \begin{bmatrix} 2 & 0 & 0 & 2 \\ 2 & 0 & 0 & 0 \\ k_1 - k_2 & -1 & 1 & k_1 \\ -(k_1 - k_2) & 1 & 1 & -k_1 \end{bmatrix} \begin{bmatrix} 1 & 0 \\ 0 & 0 \\ 0 & 0 \\ 0 & 0 \end{bmatrix} = \frac{1}{2} \begin{bmatrix} 2 & 0 \\ 2 & 0 \\ k_1 - k_2 & 0 \\ -(k_1 - k_2) & 0 \end{bmatrix}.$$

Finally,

$$F(\lambda) = \begin{bmatrix} \lambda^2 & \lambda \\ 0 & 0 \\ 0 & 0 \\ 0 & 0 \end{bmatrix} - \tilde{R}(\lambda) H = \frac{1}{2} \begin{bmatrix} -2c_1\lambda - k_1 - k_2 & 2\lambda \\ 0 & 0 \\ 0 & 0 \\ 0 & 0 \end{bmatrix}.$$

The matrix of F is, therefore,

$$F = \frac{1}{2} \begin{bmatrix} -2c_1 & 2 \\ -k_1 - k_2 & 0 \end{bmatrix}.$$

The resulting minimal P representation is:

$$\begin{bmatrix} \dot{\xi}_1 \\ \dot{\xi}_2 \end{bmatrix} = \frac{1}{2} \begin{bmatrix} -2c_1 & 2 \\ -k_1 - k_2 & 0 \end{bmatrix} \begin{bmatrix} \xi_1 \\ \xi_2 \end{bmatrix}$$

$$\begin{bmatrix} \boldsymbol{q}_1 \\ \boldsymbol{q}_2 \\ \boldsymbol{F}_1 \\ \boldsymbol{F}_2 \end{bmatrix} = \frac{1}{2} \begin{bmatrix} 2 & 0 \\ 2 & 0 \\ k_1 - k_2 & 0 \\ -(k_1 - k_2) & 0 \end{bmatrix} \begin{bmatrix} \xi_1 \\ \xi_2 \end{bmatrix}.$$

CASE II: $c_1 \neq c_2$

The matrix

$$U(s) = \begin{bmatrix} s & -s-c_1+c_2 & 0 & -s^3-c_1s^2 \\ 1 & -1 & 0 & -s^2 \\ 0 & 0 & 1 & 0 \\ 0 & 0 & 0 & 1 \end{bmatrix}$$

is unimodular and has the property that $\tilde{R}(s) = U(s)R(s)$ is row reduced with row degrees 1, 1, 0, and 0:

$$\begin{aligned} \tilde{R}(s) &= \begin{bmatrix} k_1 s & (c_2-c_1)(c_2s+k_2)-k_2s & -s & c_1-c_2+s \\ c_1s+k_1 & -c_2s-k_2 & -1 & 1 \\ 0 & 0 & 1 & 1 \\ 1 & -1 & 0 & 0 \end{bmatrix} \\ &= \operatorname{diag}(s,s,1,1)B(s), \end{aligned}$$

where

$$B(\infty) = \begin{bmatrix} k_1 & (c_2-c_1)c_2-k_2 & -1 & 1 \\ c_1 & -c_2 & 0 & 0 \\ 0 & 0 & 1 & 1 \\ 1 & -1 & 0 & 0 \end{bmatrix}$$

so that

$$B(\infty)^{-1} = \frac{1}{2a}\begin{bmatrix} 0 & -2 & 0 & 2c_2 \\ 0 & -2 & 0 & 2c_1 \\ -a & -ac_2+b & a & ac_1c_2+c_2k_1-c_1k_2 \\ a & ac_2-b & a & -ac_1c_2-c_2k_1+c_1k_2 \end{bmatrix},$$

where $a = c_2 - c_1$ and $b = k_2 - k_1$. As in Case I the matrix G is the two by two identity matrix. A polynomial basis matrix for $X_-(\tilde{R})$ is given by

$$\begin{bmatrix} 1 & 0 \\ 0 & 1 \\ 0 & 0 \\ 0 & 0 \end{bmatrix}.$$

We now have

$$H = B(\infty)^{-1}\begin{bmatrix} 1 & 0 \\ 0 & 1 \\ 0 & 0 \\ 0 & 0 \end{bmatrix} = \frac{1}{2a}\begin{bmatrix} 0 & -2 \\ 0 & -2 \\ -a & -ac_2+b \\ a & ac_2-b \end{bmatrix}.$$

Finally,

$$F(\lambda) = \begin{bmatrix} \lambda & 0 \\ 0 & \lambda \\ 0 & 0 \\ 0 & 0 \end{bmatrix} - \tilde{R}(\lambda)H = \frac{1}{2}\begin{bmatrix} c_2-c_1 & c_2^2-c_1c_2+k_1+k_2 \\ -2 & -2c_2 \\ 0 & 0 \\ 0 & 0 \end{bmatrix}.$$

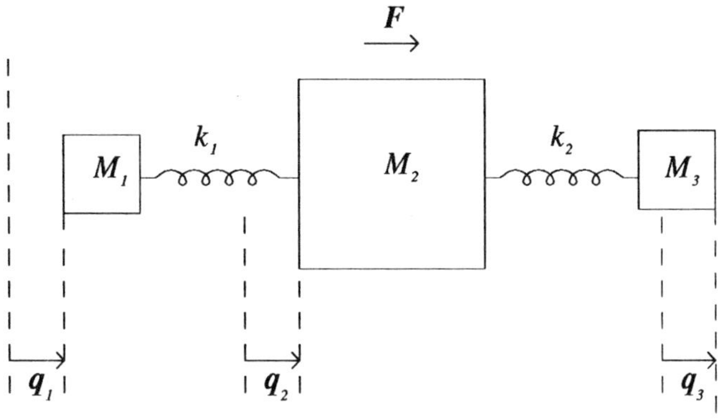

Figure 5.2: Simplified model of a satellite with flexible appendages

The matrix of F is, therefore,

$$F = \frac{1}{2}\begin{bmatrix} c_2 - c_1 & c_2^2 - c_1c_2 + k_1 + k_2 \\ -2 & -2c_2 \end{bmatrix}.$$

The resulting minimal P representation is:

$$\begin{bmatrix} \dot{\xi}_1 \\ \dot{\xi}_2 \end{bmatrix} = \frac{1}{2}\begin{bmatrix} c_2 - c_1 & c_2^2 - c_1c_2 + k_1 + k_2 \\ -2 & -2c_2 \end{bmatrix}\begin{bmatrix} \xi_1 \\ \xi_2 \end{bmatrix}$$

$$\begin{bmatrix} \boldsymbol{q}_1 \\ \boldsymbol{q}_2 \\ \boldsymbol{F}_1 \\ \boldsymbol{F}_2 \end{bmatrix} = \frac{1}{2(c_2 - c_1)}\begin{bmatrix} 0 & -2 \\ 0 & -2 \\ -(c_2 - c_1) & -(c_2 - c_1)c_2 + k_2 - k_1 \\ c_2 - c_1 & (c_2 - c_1)c_2 - k_2 + k_1 \end{bmatrix}\begin{bmatrix} \xi_1 \\ \xi_2 \end{bmatrix}.$$

Example 5.14 Consider the simplified model of a satellite with flexible appendages that is depicted in Figure 5.2. It consists of three masses that are connected by two springs and that may move along a line. The positions $\boldsymbol{q}_1$, $\boldsymbol{q}_2$ and $\boldsymbol{q}_3$ are given with respect to three fixed points in such a way that for $i = 1, 3$, the mass M_i is in equilibrium with respect to the mass M_2 if and only if $\boldsymbol{q}_i = \boldsymbol{q}_2$. With $\boldsymbol{q}_1$, $\boldsymbol{q}_2$, $\boldsymbol{q}_3$ and $\boldsymbol{F}$ as external variables, an AR representation of the system is given by

$$R(\frac{d}{dt})\begin{bmatrix} \boldsymbol{q}_1 \\ \boldsymbol{q}_2 \\ \boldsymbol{q}_3 \\ \boldsymbol{F} \end{bmatrix} = 0,$$

where

$$R(s) = \begin{bmatrix} M_1s^2 + k_1 & -k_1 & 0 & 0 \\ -k_1 & M_2s^2 + k_1 + k_2 & -k_2 & -1 \\ 0 & -k_2 & M_3s^2 + k_2 & 0 \end{bmatrix}.$$

For the sake of simplicity, we will assume that $M_1 = M_3 = 1$ and that $k_1 = k_2 = 1$. The matrix $R(s)$ is row reduced with row degrees 2, 2 and 2:

$$R(s) = \text{diag}\,(s^2, s^2, s^2)B(s),$$

where

$$B(\infty) = \begin{bmatrix} 1 & 0 & 0 & 0 \\ 0 & M_2 & 0 & 0 \\ 0 & 0 & 1 & 0 \end{bmatrix}.$$

Since dim ker $R(s) = q - r = 4 - 3 = 1$, we get $G = [I \quad 0] \in \mathbb{R}^{6\times 7}$. A polynomial basis matrix for $X_-(R)$ is given by

$$\begin{bmatrix} \lambda & 1 & 0 & 0 & 0 & 0 \\ 0 & 0 & \lambda & 1 & 0 & 0 \\ 0 & 0 & 0 & 0 & \lambda & 1 \end{bmatrix}.$$

We now have to choose $\tilde{B}$ to complete $B(\infty)$ to a nonsingular matrix; we choose

$$\tilde{B} = [\, 0 \quad 0 \quad 0 \quad 1 \,],$$

which gives

$$\hat{B}(\infty) = \begin{bmatrix} 1 & 0 & 0 & 0 \\ 0 & M_2 & 0 & 0 \\ 0 & 0 & 1 & 0 \\ 0 & 0 & 0 & 1 \end{bmatrix}$$

so that

$$\hat{B}(\infty)^{-1} = \frac{1}{M_2}\begin{bmatrix} M_2 & 0 & 0 & 0 \\ 0 & 1 & 0 & 0 \\ 0 & 0 & M_2 & 0 \\ 0 & 0 & 0 & M_2 \end{bmatrix}.$$

Therefore,

$$\begin{aligned} H &= \hat{B}(\infty)^{-1}\begin{bmatrix} 1 & 0 & 0 & 0 & 0 & 0 & 0 \\ 0 & 0 & 1 & 0 & 0 & 0 & 0 \\ 0 & 0 & 0 & 0 & 1 & 0 & 0 \\ 0 & 0 & 0 & 0 & 0 & 0 & 1 \end{bmatrix} \\ &= \frac{1}{M_2}\begin{bmatrix} M_2 & 0 & 0 & 0 & 0 & 0 & 0 \\ 0 & 0 & 1 & 0 & 0 & 0 & 0 \\ 0 & 0 & 0 & 0 & M_2 & 0 & 0 \\ 0 & 0 & 0 & 0 & 0 & 0 & M_2 \end{bmatrix}. \end{aligned}$$

Finally,

$$F(\lambda) = \begin{bmatrix} \lambda^2 & \lambda & 0 & 0 & 0 & 0 & 0 \\ 0 & 0 & \lambda^2 & \lambda & 0 & 0 & 0 \\ 0 & 0 & 0 & 0 & \lambda^2 & \lambda & 0 \end{bmatrix} - R(\lambda)H$$
$$= \frac{1}{M_2}\begin{bmatrix} -M_2 & M_2\lambda & 1 & 0 & 0 & 0 & 0 \\ M_2 & 0 & -2 & M_2\lambda & M_2 & 0 & M_2 \\ 0 & 0 & 1 & 0 & -M_2 & M_2\lambda & 0 \end{bmatrix}.$$

The matrix of F is, therefore, given by

$$F = \frac{1}{M_2}\begin{bmatrix} 0 & M_2 & 0 & 0 & 0 & 0 & 0 \\ -M_2 & 0 & 1 & 0 & 0 & 0 & 0 \\ 0 & 0 & 0 & M_2 & 0 & 0 & 0 \\ M_2 & 0 & -2 & 0 & M_2 & 0 & M_2 \\ 0 & 0 & 0 & 0 & 0 & M_2 & 0 \\ 0 & 0 & 1 & 0 & -M_2 & 0 & 0 \end{bmatrix}.$$

The resulting minimal P representation is:

$$\begin{bmatrix} \dot{\xi}_1 \\ \dot{\xi}_2 \\ \dot{\xi}_3 \\ \dot{\xi}_4 \\ \dot{\xi}_5 \\ \dot{\xi}_6 \end{bmatrix} = \frac{1}{M_2}\begin{bmatrix} 0 & M_2 & 0 & 0 & 0 & 0 & 0 \\ -M_2 & 0 & 1 & 0 & 0 & 0 & 0 \\ 0 & 0 & 0 & M_2 & 0 & 0 & 0 \\ M_2 & 0 & -2 & 0 & M_2 & 0 & M_2 \\ 0 & 0 & 0 & 0 & 0 & M_2 & 0 \\ 0 & 0 & 1 & 0 & -M_2 & 0 & 0 \end{bmatrix}\begin{bmatrix} \xi_1 \\ \xi_2 \\ \xi_3 \\ \xi_4 \\ \xi_5 \\ \xi_6 \\ \xi_7 \end{bmatrix}$$

$$\begin{bmatrix} \mathbf{q}_1 \\ \mathbf{q}_2 \\ \mathbf{q}_3 \\ \mathbf{F} \end{bmatrix} = \frac{1}{M_2}\begin{bmatrix} M_2 & 0 & 0 & 0 & 0 & 0 & 0 \\ 0 & 0 & 1 & 0 & 0 & 0 & 0 \\ 0 & 0 & 0 & 0 & M_2 & 0 & 0 \\ 0 & 0 & 0 & 0 & 0 & 0 & M_2 \end{bmatrix}\begin{bmatrix} \xi_1 \\ \xi_2 \\ \xi_3 \\ \xi_4 \\ \xi_5 \\ \xi_6 \\ \xi_7 \end{bmatrix}. \quad (5.23)$$

We remark that the system is not controllable: the matrix $R(s)$ has zeros at $s = i$ and at $s = -i$. An MA representation of the controllable part of the behavior is given by

$$\begin{bmatrix} \mathbf{q}_1 \\ \mathbf{q}_2 \\ \mathbf{q}_3 \\ \mathbf{F} \end{bmatrix} = \begin{bmatrix} 1 \\ \frac{d^2}{dt^2} + 1 \\ 1 \\ M_2\frac{d^4}{dt^4} + (M_2+2)\frac{d^2}{dt^2} \end{bmatrix}\boldsymbol{\ell}. \quad (5.24)$$

Note that the controllable subbehavior consists of those trajectories of the behavior in which the appendages move symmetrically. By using the simple realization method given at the beginning of this chapter, we get the following minimal P representation of the controllable subbehavior:

$$\begin{bmatrix} \dot{\xi}_1 \\ \dot{\xi}_2 \\ \dot{\xi}_3 \\ \dot{\xi}_4 \end{bmatrix} = \begin{bmatrix} 0 & 1 & 0 & 0 & 0 \\ 0 & 0 & 1 & 0 & 0 \\ 0 & 0 & 0 & 1 & 0 \\ 0 & 0 & 0 & 0 & 1 \end{bmatrix} \begin{bmatrix} \xi_1 \\ \xi_2 \\ \xi_3 \\ \xi_4 \\ \xi_5 \end{bmatrix}$$

$$\begin{bmatrix} \boldsymbol{q}_1 \\ \boldsymbol{q}_2 \\ \boldsymbol{q}_3 \\ \boldsymbol{F} \end{bmatrix} = \begin{bmatrix} 1 & 0 & 0 & 0 & 0 \\ 1 & 0 & 1 & 0 & 0 \\ 1 & 0 & 0 & 0 & 0 \\ 0 & 0 & M_2+2 & 0 & M_2 \end{bmatrix} \begin{bmatrix} \xi_1 \\ \xi_2 \\ \xi_3 \\ \xi_4 \\ \xi_5 \end{bmatrix}. \tag{5.25}$$

We see that we arrive at the realization of the controllable part of the behavior in a fairly easy manner (essentially due to the fact the $R(s)$ has $q-1$ rows). ◇

Let us now present a theoretical example to illustrate Procedure 5.9 in more detail. We will come back to this example in subsequent sections.

Example 5.15 Let $R(s)$ be given by

$$R(s) = \begin{bmatrix} s+1 & 0 & s^2 & 2 & 0 \\ 0 & s & 1 & s-1 & 3s \end{bmatrix}.$$

Then $R(s)$ is row reduced and, with $B(s)$ defined by (5.22),

$$B(\infty) = \begin{bmatrix} 0 & 0 & 1 & 0 & 0 \\ 0 & 1 & 0 & 1 & 3 \end{bmatrix}.$$

The row degrees are 2 and 1, and a polynomial basis matrix for $X_-(R)$ is given by

$$\begin{bmatrix} \lambda & 1 & 0 \\ 0 & 0 & 1 \end{bmatrix}.$$

Furthermore, dim ker $R(s) = 3$, so we get $G = [I \quad 0] \in \mathbb{R}^{3\times 6}$. We now have to choose $\tilde{B}$ to complete $B(\infty)$ to a nonsingular matrix; we choose

$$\tilde{B} = \begin{bmatrix} 1 & 0 & 0 & 0 & 0 \\ 0 & 0 & 0 & 1 & 0 \\ 0 & 0 & 0 & 0 & 1 \end{bmatrix},$$

which gives

$$\hat{B}(\infty) = \begin{bmatrix} 0 & 0 & 1 & 0 & 0 \\ 0 & 1 & 0 & 1 & 3 \\ 1 & 0 & 0 & 0 & 0 \\ 0 & 0 & 0 & 1 & 0 \\ 0 & 0 & 0 & 0 & 1 \end{bmatrix}$$

so that

$$\hat{B}(\infty)^{-1} = \begin{bmatrix} 0 & 0 & 1 & 0 & 0 \\ 0 & 1 & 0 & -1 & -3 \\ 1 & 0 & 0 & 0 & 0 \\ 0 & 0 & 0 & 1 & 0 \\ 0 & 0 & 0 & 0 & 1 \end{bmatrix}.$$

Therefore,

$$H = \begin{bmatrix} 0 & 0 & 1 & 0 & 0 \\ 0 & 1 & 0 & -1 & -3 \\ 1 & 0 & 0 & 0 & 0 \\ 0 & 0 & 0 & 1 & 0 \\ 0 & 0 & 0 & 0 & 1 \end{bmatrix} \begin{bmatrix} 1 & 0 & 0 & 0 & 0 & 0 \\ 0 & 0 & 1 & 0 & 0 & 0 \\ 0 & 0 & 0 & 1 & 0 & 0 \\ 0 & 0 & 0 & 0 & 1 & 0 \\ 0 & 0 & 0 & 0 & 0 & 1 \end{bmatrix} = \begin{bmatrix} 0 & 0 & 0 & 1 & 0 & 0 \\ 0 & 0 & 1 & 0 & -1 & -3 \\ 1 & 0 & 0 & 0 & 0 & 0 \\ 0 & 0 & 0 & 0 & 1 & 0 \\ 0 & 0 & 0 & 0 & 0 & 1 \end{bmatrix}.$$

Finally,

$$\begin{aligned} F(\lambda) &= \begin{bmatrix} \lambda^2 & \lambda & 0 & 0 & 0 & 0 \\ 0 & 0 & \lambda & 0 & 0 & 0 \end{bmatrix} - R(\lambda)H \\ &= \begin{bmatrix} 0 & \lambda & 0 & -\lambda-1 & -2 & 0 \\ -1 & 0 & 0 & 0 & 1 & 0 \end{bmatrix}. \end{aligned}$$

The matrix of F is, therefore,

$$F = \begin{bmatrix} 0 & 1 & 0 & -1 & 0 & 0 \\ 0 & 0 & 0 & -1 & -2 & 0 \\ -1 & 0 & 0 & 0 & 1 & 0 \end{bmatrix}.$$

The resulting minimal P representation is:

$$\begin{aligned} \begin{bmatrix} \sigma\xi_1 \\ \sigma\xi_2 \\ \sigma\xi_3 \end{bmatrix} &= \begin{bmatrix} 0 & 1 & 0 & -1 & 0 & 0 \\ 0 & 0 & 0 & -1 & -2 & 0 \\ -1 & 0 & 0 & 0 & 1 & 0 \end{bmatrix} \begin{bmatrix} \xi_1 \\ \xi_2 \\ \xi_3 \\ \xi_4 \\ \xi_5 \\ \xi_6 \end{bmatrix} \\ \begin{bmatrix} w_1 \\ w_2 \\ w_3 \\ w_4 \\ w_5 \end{bmatrix} &= \begin{bmatrix} 0 & 0 & 0 & 1 & 0 & 0 \\ 0 & 0 & 1 & 0 & -1 & -3 \\ 1 & 0 & 0 & 0 & 0 & 0 \\ 0 & 0 & 0 & 0 & 1 & 0 \\ 0 & 0 & 0 & 0 & 0 & 1 \end{bmatrix} \begin{bmatrix} \xi_1 \\ \xi_2 \\ \xi_3 \\ \xi_4 \\ \xi_5 \\ \xi_6 \end{bmatrix}. \end{aligned}$$

5.3.1 Realization in dual pencil form

In this subsection we show how a minimal DP representation can be obtained by a small modification of Procedure 5.9. The modification concerns the choice of the matrix $\tilde{B}(s)$; we will choose $\tilde{B}(s)$ such that the P representation that results from Procedure 5.9 can be trivially rewritten in dual pencil form. Details are given in the next procedure.

Procedure 5.16 As in procedure 5.9 we first factorize the matrix $R(s)$ as

$$R(s) = \Delta(s)B(s) \text{ with } \Delta(s) = \operatorname{diag}(s^{\kappa_1}, \dots, s^{\kappa_r}),$$

where $B(s)$ is right $\mathbb{R}_\infty(s)$-unimodular; as before the κ_i's coincide with the observability indices of Σ and $\dim W^0 = q - r$. The last equality implies that rank $B(\infty) = r$. By renumbering the components of w if necessary, we may assume that the first r columns of $B(\infty)$ are linearly independent. We may then choose $\tilde{B}(s)$ in Procedure 5.9 as the constant matrix $\tilde{B}$ given by

$$\tilde{B} = [0 \quad I],$$

where the partitioning is $(q-r) \times (r + (q-r))$. The matrix $\hat{B}^{-1}(\infty)$ in Procedure 5.9 is then necessarily of the form

$$\begin{bmatrix} * & * \\ 0 & I \end{bmatrix},$$

where the partitioning is $(r + (q-r)) \times (r + (q-r))$. Consequently, Procedure 5.9 leads to pencil matrices F, G and H of the form

$$\left[\begin{array}{c} sG - F \\ \hline H \end{array}\right] = \left[\begin{array}{cc} sI - F_1 & -F_2 \\ \hline H_1 & H_2 \\ 0 & I \end{array}\right],$$

where the partitioning is $(\operatorname{ord}(\Sigma) + r + (q-r)) \times (\operatorname{ord}(\Sigma) + (q-r))$. The P representation (F, G, H) is therefore written as

$$\begin{aligned} \sigma\xi_1 &= F_1\xi_1 + F_2\xi_2 \\ w &= \begin{bmatrix} H_1 \\ 0 \end{bmatrix}\xi_1 + \begin{bmatrix} H_2 \\ I_{(q-r)\times(q-r)} \end{bmatrix}\xi_2. \end{aligned}$$

This can obviously be rewritten as the DP representation (K, L, M) with

$$K = \begin{bmatrix} I \\ 0 \end{bmatrix}, \quad L = \begin{bmatrix} F_1 \\ H_1 \end{bmatrix}, \quad M = \begin{bmatrix} 0 & F_2 \\ -I_{r\times r} & H_2 \end{bmatrix}.$$

Theorem 5.17 Let $R(s)$ be a polynomial matrix of size $r \times q$ that is row reduced. Let K, L and M be defined as in Procedure 5.16. Then the DP representation (K, L, M) is strongly externally equivalent to the AR representation

$$R(\sigma)w = 0.$$

Moreover, the representation (K, L, M) is minimal under strong external equivalence, whereas it is minimal under weak external equivalence if and only if $R(s)$ has full row rank for all $s \in \mathbb{C}$.

Proof The strong external equivalence of the two representations follows immediately from the fact that (F, G, H) in Procedure 5.16 is strongly externally equivalent to the AR representation (Theorem 5.10). By construction we have that

$$\text{rank } K = \text{ord}(\Sigma) \quad \text{and} \quad \text{codim} \;\; \text{im } K = r = \text{codim } W^0$$

so that the statements on minimality follow as in the proof of Theorem 5.1; use Theorem 4.32 instead of Theorem 4.3 and Theorem 4.33 instead of Theorem 4.7.

Example 5.18 Consider the simplified model of a satellite with flexible appendages of Example 5.14. Since the first three columns of the matrix $B(\infty)$ in Example 5.14 are linearly independent, there is no need for renumbering the components of $\boldsymbol{w}$. The P representation that is obtained in Example 5.14 has the right form to apply the last step of Procedure 5.16: it can be rewritten as the minimal DP representation (K, L, M) with

$$K = \begin{bmatrix} 1&0&0&0&0&0\\ 0&1&0&0&0&0\\ 0&0&1&0&0&0\\ 0&0&0&1&0&0\\ 0&0&0&0&1&0\\ 0&0&0&0&0&1\\ 0&0&0&0&0&0\\ 0&0&0&0&0&0\\ 0&0&0&0&0&0 \end{bmatrix}, \quad L = \frac{1}{M_2}\begin{bmatrix} 0&M_2&0&0&0&0\\ -M_2&0&1&0&0&0\\ 0&0&0&M_2&0&0\\ M_2&0&-2&0&M_2&0\\ 0&0&0&0&0&M_2\\ 0&0&1&0&-M_2&0\\ M_2&0&0&0&0&0\\ 0&0&1&0&0&0\\ 0&0&0&0&M_2&0 \end{bmatrix}$$

$$M = \begin{bmatrix} 0&0&0&0\\ 0&0&0&0\\ 0&0&0&0\\ 0&0&0&1\\ 0&0&0&0\\ 0&0&0&0\\ -1&0&0&0\\ 0&-1&0&0\\ 0&0&-1&0 \end{bmatrix}.$$

Example 5.19 Let $R(s)$ be given as in Example 5.15. The first two columns of the matrix $B(\infty)$ in Example 5.15 are not linearly independent. Let us therefore interchange the components $\boldsymbol{w}_1$ and $\boldsymbol{w}_3$ and choose $\tilde{B}$ as in Procedure 5.16. It is easily verified that Procedure 5.9 then leads to the P representation

$$\begin{bmatrix} \sigma\xi_1 \\ \sigma\xi_2 \\ \sigma\xi_3 \end{bmatrix} = \begin{bmatrix} 0 & 1 & 0 & -1 & 0 & 0 \\ 0 & 0 & 0 & -1 & -2 & 0 \\ -1 & 0 & 0 & 0 & 1 & 0 \end{bmatrix} \begin{bmatrix} \xi_1 \\ \xi_2 \\ \xi_3 \\ \xi_4 \\ \xi_5 \\ \xi_6 \end{bmatrix}$$

$$\begin{bmatrix} w_3 \\ w_2 \\ w_1 \\ w_4 \\ w_5 \end{bmatrix} = \begin{bmatrix} 1 & 0 & 0 & 0 & 0 & 0 \\ 0 & 0 & 1 & 0 & -1 & -3 \\ 0 & 0 & 0 & 1 & 0 & 0 \\ 0 & 0 & 0 & 0 & 1 & 0 \\ 0 & 0 & 0 & 0 & 0 & 1 \end{bmatrix} \begin{bmatrix} \xi_1 \\ \xi_2 \\ \xi_3 \\ \xi_4 \\ \xi_5 \\ \xi_6 \end{bmatrix}.$$

Rewriting this in DP form in conformity with Procedure 5.16, we get the representation

$$\begin{bmatrix} \sigma\xi_1 \\ \sigma\xi_2 \\ \sigma\xi_3 \\ 0 \\ 0 \end{bmatrix} = \begin{bmatrix} 0 & 1 & 0 \\ 0 & 0 & 0 \\ -1 & 0 & 0 \\ 1 & 0 & 0 \\ 0 & 0 & 1 \end{bmatrix} \begin{bmatrix} \xi_1 \\ \xi_2 \\ \xi_3 \end{bmatrix} + \begin{bmatrix} 0 & 0 & -1 & 0 & 0 \\ 0 & 0 & -1 & -2 & 0 \\ 0 & 0 & 0 & 1 & 0 \\ -1 & 0 & 0 & 0 & 0 \\ 0 & -1 & 0 & -1 & -3 \end{bmatrix} \begin{bmatrix} w_3 \\ w_2 \\ w_1 \\ w_4 \\ w_5 \end{bmatrix}.$$

Consequently, the DP realization (K, L, M) of Procedure 5.16 is given by

$$K = \begin{bmatrix} 1 & 0 & 0 \\ 0 & 1 & 0 \\ 0 & 0 & 1 \\ 0 & 0 & 0 \\ 0 & 0 & 0 \end{bmatrix}, \quad L = \begin{bmatrix} 0 & 1 & 0 \\ 0 & 0 & 0 \\ -1 & 0 & 0 \\ 1 & 0 & 0 \\ 0 & 0 & 1 \end{bmatrix}, \quad M = \begin{bmatrix} -1 & 0 & 0 & 0 & 0 \\ -1 & 0 & 0 & -2 & 0 \\ 0 & 0 & 0 & 1 & 0 \\ 0 & 0 & -1 & 0 & 0 \\ 0 & -1 & 0 & -1 & -3 \end{bmatrix}.$$

◇

Our basic realization method Procedure 5.9 transforms an AR representation to pencil form. When starting from an MA representation instead of an AR representation, we can obtain the dual pencil form by a duality argument. We have the following theorem.

Theorem 5.20 Let $Q(s)$ be a polynomial matrix of size $q \times r$ that is column reduced. Apply Procedure 5.9 to the matrix $Q^T(s)$ leading to pencil matrices F, G and H. Let K, L and M be defined as $K := G^T$, $L := F^T$, and $M := H^T$. Then the DP representation (K, L, M) is weakly externally equivalent to the MA representation

$$w = Q(\sigma)\xi.$$

The representation (K, L, M) is minimal under weak external equivalence if and only if $Q(s)$ has full column rank for all $s \in \mathbb{C}$.

Proof The weak external equivalence of the two representations follows from Lemma 3.26 since (Theorem 5.10) (F, G, H) is weakly externally equivalent to the AR representation

$$Q^T(\sigma)w = 0.$$

It is obvious that (K, L, M) is minimal under weak external equivalence if and only if (F, G, H) is minimal under weak external equivalence. Hence the last statement of the theorem follows from the last statement of Theorem 5.10.

5.3.2 Realization in descriptor form

In this subsection we consider the situation in which the external variables are split into two parts. As before we use the notation $W = Y \oplus U$, where $Y = \mathbb{R}^p$ and $U = \mathbb{R}^m$. An AR representation is then written as

$$[R_1(\sigma) \quad R_2(\sigma)] \begin{bmatrix} y \\ u \end{bmatrix} = 0.$$

We will show how Procedure 5.9 can be modified to obtain minimal D representations and minimal DZ representations. As in the previous subsection the modification concerns the choice of the matrix $\tilde{B}(s)$; we will choose $\tilde{B}(s)$ such that the P representation that results from Procedure 5.9 can be trivially rewritten in D form (DZ form). Before presenting our computational procedures we first state a lemma.

Lemma 5.21 Let $\Sigma = (\mathbf{T}, W = Y \oplus U, \mathcal{B}) \in \mathcal{L}^{p+m}$ and let $[R_1(s) \quad R_2(s)]$ be a row reduced polynomial matrix of size $r \times (p + m)$ that corresponds to an AR representation of $\mathcal{B}$. Let $[R_1(s) \quad R_2(s)]$ be factorized as

$$[R_1(s) \quad R_2(s)] = \Delta(s)\,[B_1(s) \quad B_2(s)] \text{ with } \Delta(s) = \text{diag}\,(s^{\kappa_1}, \ldots, s^{\kappa_r}),$$

where $[B_1(s) \quad B_2(s)]$ is right $\mathbb{R}_\infty(s)$-unimodular and the κ_i's are the row degrees of $[R_1(s) \quad R_2(s)]$. Let W^0 be the DV space of Σ. Then the following equalities hold:

(i) $\dim\,(Y \cap W^0) = \dim\ \ker B_1(\infty)$

(ii) $\text{codim}\,(\pi_U W^0) = r - p + \dim\ \ker B_1(\infty)$

(iii) $\dim\,(\pi_Y W^0) = p - r + \dim\ \text{im}\ B_2(\infty)$

(iv) $\text{codim}\,(U \cap W^0) = \dim\ \text{im}\ B_2(\infty)$.

Proof From the definition of W^0 it follows immediately that

$$W^0 = \ker \left[B_1(\infty) \quad B_2(\infty)\right].$$

Then (i) follows trivially. Statement (ii) follows from

$$\begin{aligned} \dim \pi_U W^0 &= \dim W^0 - \dim (Y \cap W^0) \\ &= q - r - \dim \ \ker B_1(\infty). \end{aligned}$$

By interchanging Y and U, it follows that

$$\dim (U \cap W^0) = \dim \ \ker B_2(\infty),$$

which implies (iv) and also

$$\begin{aligned} \dim \pi_Y W^0 &= q - r - \dim \ \ker B_2(\infty) \\ &= p - r + \dim \ \operatorname{im} B_2(\infty). \end{aligned}$$

This proves (iii). ◇

The next procedure gives a realization in D form.

Procedure 5.22 As in procedure 5.9 we first factorize $[R_1(s) \quad R_2(s)]$ as

$$[R_1(s) \quad R_2(s)] = \Delta(s)\,[B_1(s) \quad B_2(s)] \text{ with } \Delta(s) = \operatorname{diag}\,(s^{\kappa_1}, \ldots, s^{\kappa_r}),$$

where $[B_1(s) \quad B_2(s)]$ is right $\mathbb{R}_\infty(s)$-unimodular and the κ_i's are the observability indices of Σ. By renumbering the components of u if necessary, we may assume that

$$B_2(\infty) = \left[B_2^1(\infty) \quad B_2^2(\infty)\right],$$

where $B_2^1(\infty)$ has full column rank, and the columns of $B_2^2(\infty)$ depend linearly on those of $\left[B_1(\infty) \quad B_2^1(\infty)\right]$. Denote the number of columns of $B_2^1(\infty)$ by m_1 and the number of columns of $B_2^2(\infty)$ by m_2. By construction the matrix $\left[B_1(\infty) \quad B_2^1(\infty)\right]$ has full row rank and

$$\operatorname{codim} \ \operatorname{im} B_1(\infty) = m_1.$$

We may now choose $\tilde{B}(s)$ in Procedure 5.9 as a constant matrix $\tilde{B}$ whose last m_2 rows are in the form $[0 \quad I]$. By the construction, a basis matrix for $\ker \left[B_1(\infty) \quad B_2^1(\infty)\right]$ must be of the form $[N \quad 0]^T$. Taking these facts together, we conclude that $\hat{B}(\infty)^{-1}$ in Procedure 5.9 is necessarily of the form

$$\hat{B}(\infty)^{-1} = \begin{bmatrix} * & * & * \\ * & 0 & * \\ 0 & 0 & I \end{bmatrix},$$

where the partitioning is $(p + m_1 + m_2) \times (r + (p + m - r - m_2) + m_2)$. Application of Procedure 5.9 leads to a P representation (F, G, H_y, H_u) of the form

$$\left[\begin{array}{c} sG - F \\ \hline H_y \\ \hline H_u \end{array}\right] = \left[\begin{array}{ccc} sI - F_1 & -F_2 & -F_3 \\ \hline H_{y1} & H_{y2} & H_{y3} \\ \hline H_{11} & 0 & H_{13} \\ 0 & 0 & I \end{array}\right], \tag{5.26}$$

where the partitioning is $(\mathrm{ord}(\Sigma) + p + m_1 + m_2) \times (\mathrm{ord}(\Sigma) + (p + m - r - m_2) + m_2)$. This can obviously be rewritten as the D representation (E, A, B, C, D), where

$$\begin{aligned} E &= \begin{bmatrix} I & 0 \\ 0 & 0 \end{bmatrix}, \quad A = \begin{bmatrix} F_1 & F_2 \\ H_{11} & 0 \end{bmatrix}, \quad B = \begin{bmatrix} 0 & F_3 \\ -I & H_{13} \end{bmatrix} \\ C &= [H_{y1} \quad H_{y2}], \quad D = [0 \quad H_{y3}]. \end{aligned} \tag{5.27}$$

Note that (E, A, B, C, D) is a trivial result of application of Algorithm 3.30 to (F, G, H_y, H_u).

Theorem 5.23 Let $[R_1(s) \quad R_2(s)]$ be a polynomial matrix of size $r \times (p+m)$ that is row reduced. Let E, A, B, C and D be defined as in Procedure 5.22. Then the D representation (E, A, B, C, D) is strongly externally equivalent to the AR representation

$$R_1(\sigma)\boldsymbol{y} + R_2(\sigma)\boldsymbol{u} = 0.$$

Moreover, the representation (E, A, B, C, D) is minimal under strong external equivalence, whereas it is minimal under weak external equivalence if and only if $[R_1(s) \quad R_2(s)]$ has full row rank for all $s \in \mathbb{C}$.

Proof The strong external equivalence follows immediately from the fact that (F, G, H_y, H_u) is strongly externally equivalent to the AR representation (Theorem 5.10). By construction the representation (E, A, B, C, D) has the following properties:

(i) $\mathrm{rank}\ E = \mathrm{ord}(\Sigma)$

(ii) $\dim\ker E = p+m-r-m_2 = p-r+m_1 = p-r+\mathrm{codim\ im}\ B_1(\infty) = \dim\ \ker B_1(\infty)$

(iii) $\mathrm{codim\ im}\ E = m_1 = \mathrm{codim\ im}\ B_1(\infty) = r - p + \dim\ \ker B_1(\infty)$.

It now follows from Lemma 5.21 that $\dim\ker E = \dim\,(Y \cap W^0)$ and codim im $E = \mathrm{codim}\,(\pi_U W^0)$. Together with (i) this implies the statements on minimality; see the proof of Theorem 5.1; use Theorem 4.12 instead of Theorem 4.3 and Theorem 4.19 instead of Theorem 4.7.

Example 5.24 Consider the simplified model of a satellite with flexible appendages of Example 5.14. Let us choose $\boldsymbol{q}_1$, $\boldsymbol{q}_2$ and $\boldsymbol{q}_3$ as outputs and $\boldsymbol{F}$ as input so that

$$R_1(s) = \begin{bmatrix} s^2+1 & -1 & 0 \\ -1 & M_2 s^2 + 2 & -1 \\ 0 & -1 & s^2+1 \end{bmatrix}, \quad R_2(s) = \begin{bmatrix} 0 \\ -1 \\ 0 \end{bmatrix},$$

which implies that

$$B_1(\infty) = \begin{bmatrix} 1 & 0 & 0 \\ 0 & M_2 & 0 \\ 0 & 0 & 1 \end{bmatrix}, \quad B_2(\infty) = \begin{bmatrix} 0 \\ 0 \\ 0 \end{bmatrix}.$$

In this case we have $m_1 = 0$ and $m_2 = 1$. The P representation that is obtained in Example 5.14 has the right form to apply the last step of Procedure 5.22: it can be rewritten as the minimal standard state space representation

$$\begin{bmatrix} \dot{\xi}_1 \\ \dot{\xi}_2 \\ \dot{\xi}_3 \\ \dot{\xi}_4 \\ \dot{\xi}_5 \\ \dot{\xi}_6 \end{bmatrix} = \frac{1}{M_2} \begin{bmatrix} 0 & M_2 & 0 & 0 & 0 & 0 \\ -M_2 & 0 & 1 & 0 & 0 & 0 \\ 0 & 0 & 0 & M_2 & 0 & 0 \\ M_2 & 0 & -2 & 0 & M_2 & 0 \\ 0 & 0 & 0 & 0 & 0 & M_2 \\ 0 & 0 & 1 & 0 & -M_2 & 0 \end{bmatrix} \begin{bmatrix} \xi_1 \\ \xi_2 \\ \xi_3 \\ \xi_4 \\ \xi_5 \\ \xi_6 \end{bmatrix} + \begin{bmatrix} 0 \\ 0 \\ 0 \\ 1 \\ 0 \\ 0 \end{bmatrix} \boldsymbol{F}$$

$$\begin{bmatrix} \boldsymbol{q}_1 \\ \boldsymbol{q}_2 \\ \boldsymbol{q}_3 \end{bmatrix} = \frac{1}{M_2} \begin{bmatrix} M_2 & 0 & 0 & 0 & 0 & 0 \\ 0 & 0 & 1 & 0 & 0 & 0 \\ 0 & 0 & 0 & 0 & M_2 & 0 \\ 0 & 0 & 0 & 0 & 0 & 0 \end{bmatrix} \begin{bmatrix} \xi_1 \\ \xi_2 \\ \xi_3 \\ \xi_4 \\ \xi_5 \\ \xi_6 \end{bmatrix} + \begin{bmatrix} 0 \\ 0 \\ 0 \\ 1 \end{bmatrix} \boldsymbol{F}.$$

Example 5.25 Let $R(s)$ be given as in Example 5.15. Let us choose $p = 3$ and $m = 2$ so that

$$R_1(s) = \begin{bmatrix} s+1 & 0 & s^2 \\ 0 & s & 1 \end{bmatrix}, \quad R_2(s) = \begin{bmatrix} 2 & 0 \\ s-1 & 3s \end{bmatrix}$$

and

$$B_1(\infty) = \begin{bmatrix} 0 & 0 & 1 \\ 0 & 1 & 0 \end{bmatrix}, \quad B_2(\infty) = \begin{bmatrix} 0 & 0 \\ 1 & 3 \end{bmatrix}.$$

It has been shown in Example 5.15 that $\text{ord}(\Sigma) = 3$. Clearly dim ker $B_1(\infty) = 1$, so that the E matrix in a minimal realization (E, A, B, C, D) should have size 3×4. In the notation of Procedure 5.22 we have $m_2 = 2$. The matrix $\tilde{B}$ as chosen in Example 5.15 satisfies the requirements of Procedure 5.22. Indeed, the pencil realization that was obtained in Example 5.15 already has the desired form (5.26):

$$\sigma \begin{bmatrix} 1&0&0&0&0&0 \\ 0&1&0&0&0&0 \\ 0&0&1&0&0&0 \end{bmatrix} \begin{bmatrix} \xi_1 \\ \xi_2 \\ \xi_3 \\ \xi_4 \\ \xi_5 \\ \xi_6 \end{bmatrix} = \begin{bmatrix} 0&1&0&-1&0&0 \\ 0&0&0&-1&-2&0 \\ -1&0&0&0&1&0 \end{bmatrix} \begin{bmatrix} \xi_1 \\ \xi_2 \\ \xi_3 \\ \xi_4 \\ \xi_5 \\ \xi_6 \end{bmatrix}$$

$$\begin{bmatrix} y_1 \\ y_2 \\ y_3 \\ u_1 \\ u_2 \end{bmatrix} = \begin{bmatrix} 0&0&0&1&0&0 \\ 0&0&1&0&-1&-3 \\ 1&0&0&0&0&0 \\ 0&0&0&0&1&0 \\ 0&0&0&0&0&1 \end{bmatrix} \begin{bmatrix} \xi_1 \\ \xi_2 \\ \xi_3 \\ \xi_4 \\ \xi_5 \\ \xi_6 \end{bmatrix} \tag{5.28}$$

and can be easily rewritten in (E, A, B, C, D) form:

$$E = \begin{bmatrix} 1&0&0&0 \\ 0&1&0&0 \\ 0&0&1&0 \end{bmatrix}, \quad A = \begin{bmatrix} 0&1&0&-1 \\ 0&0&0&-1 \\ -1&0&0&0 \end{bmatrix}, \quad B = \begin{bmatrix} 0&0 \\ -2&0 \\ 1&0 \end{bmatrix}$$

$$C = \begin{bmatrix} 0&0&0&1 \\ 0&0&1&0 \\ 1&0&0&0 \end{bmatrix}, \quad D = \begin{bmatrix} 0&0 \\ -1&-3 \\ 0&0 \end{bmatrix}.$$

◇

In order to arrive at a minimal descriptor representation without direct feedthrough term, i.e., a DZ representation, we have to choose $\tilde{B}$ in Procedure 5.22 differently. Details are given by the next procedure.

Procedure 5.26 As in procedure 5.22 we first factorize $[R_1(s) \quad R_2(s)]$ as

$$[R_1(s) \quad R_2(s)] = \Delta(s)\,[B_1(s) \quad B_2(s)] \text{ with } \Delta(s) = \text{diag}\,(s^{\kappa_1}, \ldots, s^{\kappa_r}),$$

where $[B_1(s) \quad B_2(s)]$ is right $\mathbb{R}_\infty(s)$-unimodular and the κ_i's are the observability indices of Σ. By renumbering the components of u if necessary, we may assume that

$$B_2(\infty) = \left[B_2^1(\infty) \quad B_2^2(\infty)\right],$$

where $B_2^1(\infty)$ has full column rank, and the columns of $B_2^2(\infty)$ depend linearly on those of $B_2^1(\infty)$. Denote the number of columns of $B_2^1(\infty)$ by m_1 and the number of columns of $B_2^2(\infty)$ by m_2. By construction

$$\dim \ \text{im}\ B_2(\infty) = m_1.$$

We may now choose $\tilde{B}(s)$ in Procedure 5.9 as a constant matrix $\tilde{B}$ of the form

$$\tilde{B} = \begin{bmatrix} \bullet & 0 & 0 \\ 0 & 0 & I \end{bmatrix},$$

where the partitioning is $((p+m-r-m_2)+m_2) \times (p+m_1+m_2)$. The matrix $\hat{B}(\infty)^{-1}$ in Procedure 5.9 is then necessarily of the form

$$\hat{B}(\infty)^{-1} = \begin{bmatrix} * & * & 0 \\ * & * & * \\ 0 & 0 & I \end{bmatrix},$$

where the partitioning is $(p+m_1+m_2) \times (r+(p+m-r-m_2)+m_2)$. Application of Procedure 5.9 leads to a P representation (F, G, H_y, H_u) of the form

$$\begin{bmatrix} sG - F \\ \hline H_y \\ \hline H_u \end{bmatrix} = \begin{bmatrix} sI - F_1 & -F_2 & -F_3 \\ \hline H_{y1} & H_{y2} & 0 \\ \hline H_{11} & H_{12} & H_{13} \\ 0 & 0 & I \end{bmatrix}, \tag{5.29}$$

where the partitioning is $(\mathrm{ord}(\Sigma)+p+m_1+m_2) \times (\mathrm{ord}(\Sigma)+(p+m-r-m_2)+m_2)$. This can obviously be rewritten as the D representation (E, A, B, C), where

$$\begin{aligned} E &= \begin{bmatrix} I & 0 \\ 0 & 0 \end{bmatrix}, \quad A = \begin{bmatrix} F_1 & F_2 \\ H_{11} & H_{12} \end{bmatrix}, \quad B = \begin{bmatrix} 0 & F_3 \\ -I & H_{13} \end{bmatrix} \\ C &= [H_{y1} \quad H_{y2}]. \end{aligned} \tag{5.30}$$

Note that (E, A, B, C) is a trivial result of application of Algorithm 3.32 to (F, G, H_y, H_u).

Theorem 5.27 Let $[R_1(s) \quad R_2(s)]$ be a polynomial matrix of size $r \times (p+m)$ that is row reduced. Let E, A, B and C be defined as in Procedure 5.26. Then the DZ representation (E, A, B, C) is strongly externally equivalent to the AR representation

$$R_1(\sigma)\boldsymbol{y} + R_2(\sigma)\boldsymbol{u} = 0.$$

Moreover, the representation (E, A, B, C) is minimal under strong external equivalence, whereas it is minimal under weak external equivalence if and only if $[R_1(s) \quad R_2(s)]$ has full row rank for all $s \in \mathbb{C}$.

Proof The strong external equivalence follows immediately from the fact that (F, G, H_y, H_u) is strongly externally equivalent to the AR representation (Theorem 5.10). By construction the representation (E, A, B, C) has the following properties:

(i) $\text{rank } E = \text{ord}(\Sigma)$

(ii) $\dim \ker E = p + m - r - m_2 = p - r + m_1 = p - r + \dim \ \text{im } B_2(\infty)$

(iii) $\text{codim im } E = m_1 = \dim \ \text{im } B_2(\infty)$.

It now follows from Lemma 5.21 that $\dim \ker E = \dim\, (\pi_Y W^0)$ and codim im E = codim $(U \cap W^0)$. Together with (i) this implies the statements on minimality; see the proof of Theorem 5.1; use Theorem 4.25 instead of Theorem 4.3 and Theorem 4.29 instead of Theorem 4.7.

Example 5.28 Let $[R_1(s) \quad R_2(s)]$ be given as in Example 5.25. We have $\text{ord}(\Sigma) = 3$ and $\dim \text{im } B_2(\infty) = 1$ so that the E matrix in a minimal realization (E, A, B, C) should have size 4×5. In the notation of Procedure 5.26 we have $m_2 = 1$. Note that the matrix $\tilde{B}$ as chosen in Example 5.15 does *not* satisfy the requirements of the above procedure. Let us choose instead

$$\tilde{B} = \begin{bmatrix} 1 & 0 & 0 & 0 & 0 \\ 0 & 1 & 0 & 0 & 0 \\ 0 & 0 & 0 & 0 & 1 \end{bmatrix}.$$

We then get

$$\hat{B}(\infty)^{-1} = \begin{bmatrix} 0 & 0 & 1 & 0 & 0 \\ 0 & 0 & 0 & 1 & 0 \\ 1 & 0 & 0 & 0 & 0 \\ 0 & 1 & 0 & -1 & -3 \\ 0 & 0 & 0 & 0 & 1 \end{bmatrix}$$

so that

$$H = \begin{bmatrix} 0 & 0 & 1 & 0 & 0 \\ 0 & 0 & 0 & 1 & 0 \\ 1 & 0 & 0 & 0 & 0 \\ 0 & 1 & 0 & -1 & -3 \\ 0 & 0 & 0 & 0 & 1 \end{bmatrix} \begin{bmatrix} 1 & 0 & 0 & 0 & 0 & 0 \\ 0 & 0 & 1 & 0 & 0 & 0 \\ 0 & 0 & 0 & 1 & 0 & 0 \\ 0 & 0 & 0 & 0 & 1 & 0 \\ 0 & 0 & 0 & 0 & 0 & 1 \end{bmatrix} = \begin{bmatrix} 0 & 0 & 0 & 1 & 0 & 0 \\ 0 & 0 & 1 & 0 & 1 & 0 \\ 1 & 0 & 0 & 0 & 0 & 0 \\ 0 & 0 & 1 & 0 & -1 & -3 \\ 0 & 0 & 0 & 0 & 0 & 1 \end{bmatrix}.$$

Furthermore,

$$\begin{aligned} F(\lambda) &= \begin{bmatrix} \lambda^2 & \lambda & 0 & 0 & 0 & 0 \\ 0 & 0 & \lambda & 0 & 0 & 0 \end{bmatrix} - R(\lambda)H \\ &= \begin{bmatrix} 0 & \lambda & -2 & -\lambda - 1 & 2 & 6 \\ -1 & 0 & 1 & 0 & -1 & -3 \end{bmatrix}. \end{aligned}$$

The matrix of F is therefore given by

$$F = \begin{bmatrix} 0 & 1 & 0 & -1 & 0 & 0 \\ 0 & 0 & -2 & -1 & 2 & 6 \\ -1 & 0 & 1 & 0 & -1 & -3 \end{bmatrix}.$$

As in Example 5.15 we have $G = [I \quad 0]$ of size 3×6. Procedure 5.9 leads to the P realization

$$\sigma \begin{bmatrix} 1 & 0 & 0 & 0 & 0 & 0 \\ 0 & 1 & 0 & 0 & 0 & 0 \\ 0 & 0 & 1 & 0 & 0 & 0 \end{bmatrix} \begin{bmatrix} \xi_1 \\ \xi_2 \\ \xi_3 \\ \xi_4 \\ \xi_5 \\ \xi_6 \end{bmatrix} = \begin{bmatrix} 0 & 1 & 0 & -1 & 0 & 0 \\ 0 & 0 & -2 & -1 & 2 & 6 \\ -1 & 0 & 1 & 0 & -1 & -3 \end{bmatrix} \begin{bmatrix} \xi_1 \\ \xi_2 \\ \xi_3 \\ \xi_4 \\ \xi_5 \\ \xi_6 \end{bmatrix}$$

$$\begin{bmatrix} y_1 \\ y_2 \\ y_3 \\ u_1 \\ u_2 \end{bmatrix} = \begin{bmatrix} 0 & 0 & 0 & 1 & 0 & 0 \\ 0 & 0 & 0 & 0 & 1 & 0 \\ 1 & 0 & 0 & 0 & 0 & 0 \\ 0 & 0 & 1 & 0 & -1 & -3 \\ 0 & 0 & 0 & 0 & 0 & 1 \end{bmatrix} \begin{bmatrix} \xi_1 \\ \xi_2 \\ \xi_3 \\ \xi_4 \\ \xi_5 \\ \xi_6 \end{bmatrix}. \qquad (5.31)$$

This representation can be easily rewritten in (E, A, B, C) form:

$$E = \begin{bmatrix} 1 & 0 & 0 & 0 & 0 \\ 0 & 1 & 0 & 0 & 0 \\ 0 & 0 & 1 & 0 & 0 \end{bmatrix}, \quad A = \begin{bmatrix} 0 & 1 & 0 & -1 & 0 \\ 0 & 0 & -2 & -1 & 2 \\ -1 & 0 & 1 & 0 & -1 \\ 0 & 0 & 1 & 0 & 0 \end{bmatrix},$$

$$B = \begin{bmatrix} 0 & 0 \\ 0 & 6 \\ 0 & 3 \\ -1 & -3 \end{bmatrix}, \quad C = \begin{bmatrix} 0 & 0 & 0 & 1 & 0 \\ 0 & 0 & 1 & 0 & 0 \\ 1 & 0 & 0 & 0 & 0 \end{bmatrix}.$$

Remark 5.29 We conclude from the proofs of Theorem 5.23 and Theorem 5.27 that the matrix E of the descriptor representation that is obtained in Procedure 5.22/Procedure 5.26 is square if and only if $r = p$, i.e., the number of y-components equals the number of equations of the AR representation that one starts with.

5.4 Connections with the Fuhrmann realization

In this section we consider I-O systems $\Sigma = (\mathbf{T}, W = Y \oplus U, \mathcal{B}) \in \mathcal{L}^{p+m}$ as defined in Definition 3.27 for which the transfer function $T(s) \in \mathbb{R}^{p \times m}(s)$ is strictly proper. We consider the strictly proper case to compare the basis-free realization method of Section 5.1 with the realization method by Fuhrmann [16, 18]. We will see that the Fuhrmann realization can be recovered from our pencil realization. In this way we show that the Fuhrmann realization fits well into the behavioral approach.

First we recall the Fuhrmann realization method.

Procedure 5.30 (*Fuhrmann realization method*) Let $T(s)$ be a strictly proper rational matrix in $\mathbb{R}^{p\times m}(s)$. First factorize $T(s)$ as

$$T(s) = R_1^{-1}(s)R_2(s), \tag{5.32}$$

where $R_1(s)$ and $R_2(s)$ are polynomial matrices. Next define the following space of polynomial vectors

$$X_-(R_1) := \{p(\lambda) \in \mathbb{R}^p[\lambda] \mid R_1^{-1}(\lambda)p(\lambda) \text{ is strictly proper }\}.$$

Define mappings A from $X_-(R_1)$ to $X_-(R_1)$, B from U to $X_-(R_1)$, and C from $X_-(R_1)$ to Y as follows:

$$A : p(\lambda) \mapsto R_1(\lambda)\pi_-(\lambda R_1^{-1}(\lambda)p(\lambda)) \tag{5.33}$$
$$B : u \mapsto R_1(\lambda)\pi_-(R_1^{-1}(\lambda)R_2(\lambda))u \ (= R_2(\lambda)u) \tag{5.34}$$
$$C : p(\lambda) \mapsto \pi_1(R_1^{-1}(\lambda)p(\lambda)). \tag{5.35}$$

◇

As remarked before the transfer function is a complete invariant under weak external equivalence. The theorem by Fuhrmann [16, 18] can therefore be formulated in our terminology as follows:

Theorem 5.31 Let $\Sigma = (\mathbf{T}, W = Y \oplus U, \mathcal{B}) \in \mathcal{L}^{p+m}$ be an I-O system with strictly proper transfer function $T(s) = R_1^{-1}(s)R_2(s)$, where $R_1(s)$ and $R_2(s)$ are polynomial matrices. Let A, B and C be defined by (5.33–5.35). Then the DZ representation (I, A, B, C) is weakly externally equivalent to the AR representation

$$R_1(\sigma)\boldsymbol{y} + R_2(\sigma)\boldsymbol{u} = 0.$$

Proof Since the transfer function is a complete invariant under weak external equivalence, it is sufficient to prove that the transfer functions corresponding to the representations are the same, i.e., that

$$R_1^{-1}(s)R_2(s) = C(sI - A)^{-1}B. \tag{5.36}$$

From the definition of A and C, we have, for any $p(\lambda) \in X_-(R_1)$, that $(\lambda I - A)p(\lambda) = R_1(\lambda)Cp(\lambda)$. As a result $R_1(\lambda)C(\lambda I - A)^{-1}Bu = R_2(\lambda)u$ for all $u \in U$. From this it follows that (5.36) holds. ◇

The next theorem shows that the above representation (I, A, B, C) is actually a minimal representation of the *behavior* of Σ.

Theorem 5.32 Let $\Sigma = (\mathbf{T}, W = Y \oplus U, \mathcal{B}) \in \mathcal{L}^{p+m}$ be an I-O system with strictly proper transfer function $T(s) = R_1^{-1}(s)R_2(s)$, where $R_1(s)$ and $R_2(s)$ are polynomial matrices. Let A, B and C be defined by (5.33–5.35). Then the DZ representation (I, A, B, C) is strongly externally equivalent to the AR representation

$$R_1(\sigma)\boldsymbol{y} + R_2(\sigma)\boldsymbol{u} = 0. \tag{5.37}$$

Moreover, the representation (I, A, B, C) is minimal under strong external equivalence, whereas it is minimal under weak external equivalence if and only if $[R_1(s) \quad R_2(s)]$ has full row rank for all $s \in \mathbb{C}$. $\diamond$

For the proof of this theorem, we first need a lemma. Recall the definitions of $\mathcal{W}_0$, $\mathcal{W}_*$ and $X_-(R)$:

$$\mathcal{W}_0 = \{\begin{bmatrix} y(\lambda) \\ u(\lambda) \end{bmatrix} \in \lambda^{-1}W_\infty(\lambda) \mid R_1(\lambda)y(\lambda) + R_2(\lambda)u(\lambda) \text{ is polynomial} \}$$

$$\mathcal{W}_* = \{\begin{bmatrix} y(\lambda) \\ u(\lambda) \end{bmatrix} \in \lambda^{-1}W_\infty(\lambda) \mid R_1(\lambda)y(\lambda) + R_2(\lambda)u(\lambda) = 0\}$$

$$X_-(R) = \{p(\lambda) \in \mathbb{R}^p[\lambda] \mid \exists \begin{bmatrix} y(\lambda) \\ u(\lambda) \end{bmatrix} \in \mathcal{W}_0 \text{ such that}$$
$$p(\lambda) = R_1(\lambda)y(\lambda) + R_2(\lambda)u(\lambda)\}.$$

Lemma 5.33 Let $[R_1(s) \quad R_2(s)]$ be a polynomial matrix with $R_1(s)$ invertible and $R_1^{-1}(s)R_2(s)$ strictly proper. Then

$$X_-(R) = X_-(R_1)$$

and the mapping Φ from $\mathcal{W}_0/\lambda^{-1}\mathcal{W}_*$ to $X_-(R_1) \oplus U$ defined by

$$\Phi : [w(\lambda)] \mapsto \begin{bmatrix} R(\lambda)w(\lambda) \\ \pi_U(\pi_1(w(\lambda))) \end{bmatrix}$$

is an isomorphism; its inverse is given by

$$\begin{aligned} \Phi^{-1} : X_-(R_1) \oplus U &\to \mathcal{W}_0/\lambda^{-1}\mathcal{W}_* \\ \begin{bmatrix} p(\lambda) \\ u \end{bmatrix} &\mapsto [\begin{bmatrix} R_1^{-1}(\lambda)p(\lambda) - \lambda^{-1}R_1^{-1}(\lambda)R_2(\lambda)u \\ \lambda^{-1}u \end{bmatrix}]. \end{aligned}$$

Proof We first prove that $X_-(R_1) = X_-(R)$. Let $p(\lambda) \in X_-(R)$ be written as $p(\lambda) = R_1(\lambda)y(\lambda) + R_2(\lambda)u(\lambda)$ with $y(\lambda) \in \lambda^{-1}Y_\infty(\lambda)$ and $u(\lambda) \in \lambda^{-1}U_\infty(\lambda)$. Then the vector $R_1^{-1}(\lambda)p(\lambda) = y(\lambda) + R_1^{-1}(\lambda)R_2(\lambda)u(\lambda)$ is strictly proper, which proves that $X_-(R_1) \supset X_-(R)$. The inclusion $X_-(R_1) \subset X_-(R)$ is trivial.

To prove that Φ is injective, let $y(\lambda) \in \lambda^{-1}Y_\infty(\lambda)$ and $u(\lambda) \in \lambda^{-1}U_\infty(\lambda)$ be such that $R_1(\lambda)y(\lambda) + R_2(\lambda)u(\lambda) = 0$ and $\pi_1(u(\lambda)) = 0$. Since $y(\lambda) = -R_1^{-1}(\lambda)R_2(\lambda)u(\lambda)$ and $R_1^{-1}(\lambda)R_2(\lambda)$ is strictly proper, it follows that $\pi_1(y(\lambda)) = 0$. More generally, this shows that $\pi_1\mathcal{W}_* = U$, i.e., that

$$W^0 = U \tag{5.38}$$

and

$$\begin{bmatrix} y(\lambda) \\ u(\lambda) \end{bmatrix} \in \mathcal{W}_0/\lambda^{-1}\mathcal{W}_*,$$

which proves the injectivity of Φ. The invertibility of Φ can be concluded from a dimensionality argument: dim $X_-(R_1)\oplus U =$ dim $X_-(R)+$dim $U =$ dim $\mathcal{W}_0/\mathcal{W}_* +$ dim $U =$ dim $\mathcal{W}_0/\mathcal{W}_* +$ dim $W^0 =$ dim $\mathcal{W}_0/\lambda^{-1}\mathcal{W}_*$. Here we used (5.38) in the third equality. It is easy to check that the mapping $\Phi\Phi^{-1}$ is the identity. Since Φ is invertible, it then follows that Φ^{-1} is indeed the inverse of Φ.

Proof of Theorem 5.32 Let (F, G, H_y, H_u) be the pencil realization of Section 5.1; G and F are then mappings from $\mathcal{W}_0/\lambda^{-1}\mathcal{W}_*$ to $X_-(R)$ that are defined by (5.5) and (5.6):

$$G : [\begin{bmatrix} y(\lambda) \\ u(\lambda) \end{bmatrix}] \mapsto R_1(\lambda)y(\lambda) + R_2(\lambda)u(\lambda)$$

$$F : [\begin{bmatrix} y(\lambda) \\ u(\lambda) \end{bmatrix}] \mapsto R_1(\lambda)\pi_-(\lambda y(\lambda)) + R_2(\lambda)\pi_-(\lambda u(\lambda))$$

and H_y and H_u are mappings from $\mathcal{W}_0/\lambda^{-1}\mathcal{W}_*$ to Y and U, respectively, that are defined by (5.7):

$$H_y : [\begin{bmatrix} y(\lambda) \\ u(\lambda) \end{bmatrix}] \mapsto \pi_1 y(\lambda)$$

$$H_u : [\begin{bmatrix} y(\lambda) \\ u(\lambda) \end{bmatrix}] \mapsto \pi_1 u(\lambda).$$

The P representation (F, G, H_y, H_u) is strongly externally equivalent to (5.37) by Theorem 5.1. With Φ defined as in Lemma 5.33, we now define mappings $\tilde{G}$ and $\tilde{F}$ from $X_-(R_1) \oplus U$ to $X_-(R_1)$ by

$$\tilde{G} = G\Phi^{-1} \text{ and } \tilde{F} = F\Phi^{-1}$$

and mappings $\tilde{H}_y$ and $\tilde{H}_u$ from $X_-(R_1) \oplus U$ to Y and U, respectively, by

$$\tilde{H}_y = H_y\Phi^{-1} \text{ and } \tilde{H}_u = H_u\Phi^{-1}.$$

The P representation $(\tilde{F}, \tilde{G}, \tilde{H}_y, \tilde{H}_u)$ is also strongly externally to (5.37) because of Theorem 4.34 and the facts that $X_-(R) = X_-(R_1)$ and that Φ is an isomorphism (Lemma 5.33). By checking the definitions, it is easily verified that $(\tilde{F}, \tilde{G}, \tilde{H}_y, \tilde{H}_u)$ coincides with the DZ representation (I, A, B, C):

$$\tilde{G} = [I \quad 0], \quad \tilde{F} = [A \quad B], \quad \tilde{H}_y = [C \quad 0], \quad \tilde{H}_u = [0 \quad I].$$

This proves that (I, A, B, C) is strongly externally equivalent to (5.37); the statements on minimality are immediate from analogous statements in Theorem 5.1.

Remark 5.34 When a rational matrix $T(s)$, factorized as in (5.32) as

$$T(s) = R_1^{-1}(s)R_2(s), \tag{5.39}$$

is *nonproper*, we may still use the Fuhrmann realization method arriving at

$$\pi_- T(s) = C(sI - A)^{-1}B,$$

that is, a standard state space realization of the strictly proper part of $T(s)$. Observe that $R_1(s)$ and $R_2(s)$ are polynomial matrices and therefore have no finite poles. By "reflection" we may use Fuhrmann's method to obtain an expression for the *"strictly" polynomial* part $\pi_+ T(s)$ of the form (see [32])

$$\pi_+ T(s) = \tilde{C}\left(s^{-1}I - \tilde{A}\right)^{-1}\tilde{B}.$$

The construction of $\tilde{A}$, $\tilde{B}$ and $\tilde{C}$ is then based on a factorization

$$T(s) = B_1^{-1}(s)B_2(s), \tag{5.40}$$

where $B_1(s)$ and $B_2(s)$ are matrices that have no poles at infinity, i.e., are *proper* rational (note that $B_1(s)$ and $B_2(s)$ can be obtained from the left Wiener-Hopf factorization of $[R_1(s) \quad R_2(s)]$, see Theorem 2.19). The construction involves a space $X_+(B_1)$ that is the "reflection" of the state space $X_-(R_1)$:

$$X_+(B_1) = \{f(\lambda) \in \mathbb{R}^p_\infty(\lambda) \mid {B_1}^{-1}(\lambda)f(\lambda)$$
$$\text{is polynomial with zero constant term}\}.$$

The question arises: how do the spaces $X_-(R_1)$ and $X_+(B_1)$ relate to the space $X_-([R_1 \quad R_2])$ that acted as an internal variable space in the pencil realization procedure of Section 5.1? An answer is provided by [32] (see also [31]), where a mapping is constructed between $X_-(R)/X_-(R_1)$ to $X_+(B_1)$; the mapping is an isomorphism if the matrix $[B_1(s) \quad B_2(s)]$ is right $\mathbb{R}_\infty(s)$-unimodular. Under this assumption it follows in particular that

$$\dim X_-(R) = \dim X_-(R_1) + \dim X_+(B_1).$$

It is easily verified (use Remark 2.20 and Lemma 2.22) that dim $X_-(R_1) = c_\infty(\det R_1)$ and dim $X_+(B_1) = -c_\infty(\det B_1)$. Consequently, Theorem 2.15 implies that $\delta^p_\infty(T) = \dim X_+(B_1)$. When, in addition, $[R_1(s) \quad R_2(s)]$ is right $\mathbb{R}[s]$-unimodular, then Theorem 2.15 implies that

$$\sum_{\beta \in \mathbb{C}} \delta^p_\beta(T) = \dim X_-(R_1)$$

and we also have that dim $X_-(R) = \kappa^\ell(R) = $ C-ord(Σ). From this we recover (3.31):

$$\text{C-ord}(\Sigma) = \delta^p(T).$$

In fact, the above can be extended to the case in which the factorizations (5.39) and (5.40) are given with respect to general complementary subsets of $\mathbb{C}^e$ (e.g. stable/unstable; see cases (i) and (ii) in Remark 2.26). Here the key result is the fact that the Wiener-Hopf factorization can also be carried out with respect to a subset Γ of $\mathbb{C}^e$; see Remark 2.26. Expressions that can be obtained at this level of generality are of the form

$$T(s) = C(\chi(s)I - A)^{-1}B + T_0 + \tilde{C}\left(\chi^{-1}(s)I - \tilde{A}\right)^{-1}\tilde{B}$$

and give rise to expressions of the form

$$T(s) = C(\chi(s)E - A)^{-1}B.$$

Here T_0 denotes the constant part of the rational matrix $T(s)$, and $\chi(s)$ is a rational function (as in Remark 2.26) that has exactly one pole in Γ and one zero in $\mathbb{C}^e \setminus \Gamma$. The realization via the pencil form can also be performed at this level of generality, leading to an expression of the form

$$\ker\ [R_1(s) \quad R_2(s)] = H[\ \ker\ (\chi(s)G - F)].$$

These generalizations are described in detail in [32]. When working over the real numbers, we should make some additional assumptions on Γ: both Γ and $\mathbb{C}^e \setminus \Gamma$ should be symmetric with respect to the real axis, and their intersections with the extended real axis should be nonempty.

6

Structural invariants

In this chapter we are interested in a number of invariants that are related to the structure of dynamical systems. It is our aim to derive explicit formulas for the invariants in terms of the first-order representations considered before. In the next two sections we consider observability indices and controllability indices, respectively, as defined in Definition 3.11. In the last section of the chapter we study the input-output status of a dynamical system and consider the question under what conditions a system is an I-O system in the sense of Definition 3.27. Subsequently, we study certain integer invariants that are specifically related to an I-O system, namely the rank and the pole/zero structure at infinity of its transfer function. The geometric formulas that are derived in this chapter apply to representations with an arbitrary amount of redundancy. We believe this aspect to be important in practice, where systems that are an interconnection of subsystems are often modeled in a highly redundant way. By putting minor nonredundancy constraints on the representations, the expressions can be stated completely in terms of matrix pencils. For this, the Kronecker invariants of a matrix pencil (Section 2.6) play an important role. Some of the results of this chapter have been written down in [34, 35].

6.1 Observability indices

The observability indices (Definition 3.11) of a system in $\mathcal{L}^q$ are invariants that are very much related to the behavior of the system: they are invariant under strong external equivalence. In this section it is our aim to derive expressions for these indices in terms of the first-order representations that we considered previously. An important role will be played by the nonincreasing sequence $\{W_k^0\}$, where $W_k^0 \subset W$ is defined by (3.9) as

$$W_k^0 := \pi_1\{w(\lambda) \in \lambda^{-1}\mathcal{W}_\infty(\lambda) \mid \lambda^{-k}w(\lambda) \in \mathcal{W}_0 \text{ for all } k \in \mathbb{Z}_+\}. \quad (6.1)$$

The nonzero observability indices $\kappa_1, \kappa_2, \ldots, \kappa_r$ are determined by W_k^0, since equation (3.10) implies that for all $k \geq 1$

$$\sharp\{j \mid \kappa_j = k\} = \dim W_{k-1}^0 - \dim W_k^0. \quad (6.2)$$

In the following subsections we will present results for P representations, DP representations and D representations.

6.1.1 Results for P representations

In this subsection we derive expressions for the nonzero observability indices in terms of the matrices F, G and H of a P representation

$$\begin{aligned} \sigma G\boldsymbol{\xi} &= F\boldsymbol{\xi} \\ \boldsymbol{w} &= H\boldsymbol{\xi}. \end{aligned}$$

As before, G and F are mappings from Z to X, and H is a mapping from Z to W. Our strategy will be to derive an expression for W_k^0 in terms of F, G and H. Because of (6.2), formulas for the nonzero observability indices are then immediate. As a first step, we introduce the following iteration:

$$\mathcal{Z}^0 = Z, \quad \mathcal{Z}^{m+1} = F^{-1}G\mathcal{Z}^m. \tag{6.3}$$

Note that this iteration coincides with iteration (2.24) applied to the pencil $sG - F$. The limit space $\mathcal{Z}^*$ of (6.3) is related to redundancy in the representation (F, G, H): when G has full row rank, then $\mathcal{Z}^* = Z$. More precisely, Lemma 2.37 yields the following implication:

$$F[\ker G] + \operatorname{im} G = X \implies G\mathcal{Z}^* = \operatorname{im} G. \tag{6.4}$$

Theorem 6.1 Let $\Sigma = (\mathbf{T}, W, \mathcal{B}) \in \mathcal{L}^q$ and let $\mathcal{B}$ be given by a P representation (F, G, H). Let $\mathcal{Z}^*$ be the limit space of the iteration (6.3) and let $\mathcal{Q}^k \subset Z$ be given by the iteration

$$\mathcal{Q}^0 = Z, \quad \mathcal{Q}^{m+1} = G^{-1}F[\mathcal{Q}^m \cap \ker H]. \tag{6.5}$$

Denote the number of observability indices of Σ that are equal to k by π_k. Then we have, for all $k \geq 1$,

$$\pi_k = \dim H[\mathcal{Q}^{k-1} \cap \mathcal{Z}^*] - \dim H[\mathcal{Q}^k \cap \mathcal{Z}^*].$$

Proof Because of (6.2), it is obviously sufficient to prove that, for all $k \geq 0$,

$$W_k^0 = H[\mathcal{Q}^k \cap \mathcal{Z}^*].$$

For this, let $k \geq 0$ and let $w \in W_k^0$; we will first prove that $W_k^0 \subset H[\mathcal{Q}^k \cap \mathcal{Z}^*]$. By definition there exists a vector $w(\lambda) \in \lambda^{-1}W_\infty(\lambda)$ such that $\pi_1(w(\lambda)) = w$ and $\lambda^{-k}w(\lambda)$ belongs to the RB space $\mathcal{W}_0$ of Σ. It follows from Lemma 3.16 that there exists an element $\xi(\lambda) \in Z(\lambda)$ such that $(\lambda G - F)\xi(\lambda)$ is polynomial and $\lambda^{-k}w(\lambda) = \pi_-(H\xi(\lambda))$ (here π_- denotes projection onto the strictly proper part). Without loss of generality we may assume that $\xi(\lambda)$ is strictly proper so that

$$(\lambda G - F)\xi(\lambda) \text{ is a constant} \tag{6.6}$$

and

$$w(\lambda) = \lambda^k H\xi(\lambda). \tag{6.7}$$

Let us write the Laurent expansion of $\xi(\lambda)$ as

$$\xi(\lambda) = \xi_1\lambda^{-1} + \xi_2\lambda^{-2} + \cdots.$$

It follows from (6.6) that $G\xi_{i+1} = F\xi_i$ for all $i \geq 1$ so that $\xi_i \in \mathcal{Z}^*$ for all $i \geq 1$. From (6.7) it follows that $w = H\xi_{k+1}$ and that $H\xi_i = 0$ for $1 \leq i \leq k$. We conclude that $\xi_{k+1} \in \mathcal{Q}^k \cap \mathcal{Z}^*$ so that $w \in H[\mathcal{Q}^k \cap \mathcal{Z}^*]$.

Conversely, let $w = Hz$ with $z \in \mathcal{Q}^k \cap \mathcal{Z}^*$. Since $z \in \mathcal{Z}^*$, it follows from Lemma 2.29 (use Remark 2.30) that there exists a strictly proper rational function $\tilde{\xi}(\lambda)$ such that $(\lambda G - F)\tilde{\xi}(\lambda) = Fz$. Because of the fact that $z \in \mathcal{Q}^k$, we can find $\xi_i \in Z$ for $1 \leq i \leq k$ such that

$$Gz = F\xi_k,\ G\xi_k = F\xi_{k-1}, \ldots, G\xi_2 = F\xi_1$$

while

$$H\xi_k = H\xi_{k-1} = \cdots = H\xi_1 = 0.$$

Now define

$$\xi(\lambda) := \xi_1\lambda^{-1} + \xi_2\lambda^{-2} + \cdots + \xi_k\lambda^{-k} + (z + \tilde{\xi}(\lambda))\lambda^{-k-1}.$$

Then $(\lambda G - F)\tilde{\xi}(\lambda) = G\xi_1$ is polynomial, and the vector $w(\lambda)$ defined by $w(\lambda) := \lambda^k H\xi(\lambda)$ is strictly proper with $w = \pi_1 w(\lambda)$. In a trivial way we have that $\lambda^{-k}w(\lambda) - H\xi(\lambda) = 0$ is polynomial, so that it follows from Lemma 3.16 that $w \in W_k^0$. We conclude that $W_k^0 = H[\mathcal{Q}^k \cap \mathcal{Z}^*]$, which proves the theorem. ◇

Below we want to compare the geometric formula of Theorem 6.1 with expressions in terms of matrix pencils. For this, we need geometric formulas that describe the Kronecker indices of an arbitrary matrix pencil $sE - A$. Recall from Section 2.6 that the right Kronecker indices of $sE - A$ are defined as the nonzero minimal indices (see Definition 2.27) of the rational vector space ker $(sE - A)$, whereas the left Kronecker indices of $sE - A$ are defined as the nonzero minimal indices of the rational vector space ker $(sE - A)^T$. The next theorem shows that geometric formulas for arbitrary matrix pencils can be obtained from Theorem 6.1 in an elegant way.

Theorem 6.2 Let $E, A : X \to Z$ be linear mappings. Denote the number of left (right) Kronecker indices of $sE - A$ that are equal to k by q_k (r_k). Let $Q^k \subset X$ be defined by the iteration

$$Q^0 = X, \quad Q^{m+1} = E^{-1}AQ^m \tag{6.8}$$

and let $R^k \subset X$ be defined by the iteration

$$R^0 = \{0\}, \quad R^{m+1} = A^{-1}ER^m. \tag{6.9}$$

Then we have, for $k \geq 1$,

(i) $q_k = \dim\,(AQ^{k-1} + \text{ im } E) - \dim\,(AQ^k + \text{ im } E)$

(ii) $r_k = \dim\,(R^{k+1} \cap \text{ ker } E) - \dim\,(R^k \cap \text{ ker } E)$.

Proof Let us denote the number of rows and the rank of $sE - A$ by m and r, respectively. Consider the system $\Sigma = (\mathbf{T}, W, \mathcal{B}) \in \mathcal{L}_c^q$ with $W = Z$ whose behavior $\mathcal{B}$ is given by the MA representation

$$w = (\sigma E - A)\boldsymbol{\xi}. \tag{6.10}$$

According to Theorem 3.22, the observability indices $\kappa_1, \kappa_2, \ldots, \kappa_{m-r}$ of Σ coincide with the minimal indices of the space ker $(sE - A)^T$. Consequently, the nonzero observability indices coincide with the left Kronecker indices of $sE - A$. It is obvious that (6.10) can be rewritten as the P representation (F, G, H) with

$$G = [E \quad 0]\,, F = [A \quad I]\,, H = [0 \quad I]\,.$$

Let $\mathcal{Z}^*$ be the limit space of the iteration (6.3). It follows from (6.4) that

$$G\mathcal{Z}^* = \text{ im } G = \text{ im } E$$

so that $\mathcal{Z}^* = F^{-1}[\text{ im } G] = \{(x, w) \in X \oplus W \mid Ax + w \in \text{ im } E\}$. With $\mathcal{Q}^k$ given by the iteration (6.5) we have, for all $k \geq 0$,

$$\mathcal{Q}^k = Q^k \oplus W.$$

Application of Theorem 6.1 yields (i). To prove (ii), we will make use of result (i) by noting that the right Kronecker indices of $sE - A$ are the left Kronecker indices of $sE^T - A^T$. Let us now introduce the following iteration in Z

$$S^0 = \{0\}, \quad S^{m+1} = EA^{-1}S^m.$$

Using some elementary results from linear algebra, we conclude from (i) that, for $k \geq 1$,

$$r_k = \dim\,(A^{-1}S^k \cap \text{ ker } E) - \dim\,(A^{-1}S^{k-1} \cap \text{ ker } E).$$

Clearly, $R^{k+1} = A^{-1}S^k$ for $k \geq 0$ so that (ii) follows. ◇

It was already apparent from the proof of Lemma 4.2 that the nonzero observability indices of a system in pencil form (F, G, H) coincide with the left Kronecker indices of the pencil

$$\begin{bmatrix} sG - F \\ H \end{bmatrix} \tag{6.11}$$

if the matrix G has full row rank. Because of the above theorem, we may impose a slightly weaker nonredundancy condition to obtain the same result (use Lemma 2.37 to see that the condition is indeed weaker).

Corollary 6.3 Let $\Sigma = (\mathbf{T}, W, \mathcal{B}) \in \mathcal{L}^q$ and let $\mathcal{B}$ be given by a P representation (F, G, H) for which

$$sG - F \text{ has full row rank and has no zeros at infinity.} \tag{6.12}$$

Then the nonzero observability indices of Σ coincide with the left Kronecker indices of the pencil (6.11).

Proof Let $\mathcal{Z}^*$, $\mathcal{Q}^k$ and π_k be defined as in Theorem 6.1. By Lemma 2.37 the condition (6.12) is equivalent to the following condition:

$$F[\ \ker G] + \text{ im } G = X. \tag{6.13}$$

By the same lemma it follows from (6.12) that $G\mathcal{Z}^* = \text{ im } G$ and in particular that $\mathcal{Z}^* = F^{-1}[\text{ im } G]$. Consequently, Theorem 6.1 yields, for all $k \geq 1$,

$$\pi_k = \dim H[\mathcal{Q}^{k-1} \cap F^{-1}[\text{ im } G]] - \dim H[\mathcal{Q}^k \cap F^{-1}[\text{ im } G]].$$

Since $\ker G \subset \mathcal{Q}^k$ for all $k \geq 0$, it follows from (6.13) that, for all $k \geq 0$,

$$F\mathcal{Q}^k + \text{ im } G = X.$$

This implies that

$$\dim \left(\begin{bmatrix} F \\ H \end{bmatrix} \mathcal{Q}^k + \text{ im } \begin{bmatrix} G \\ 0 \end{bmatrix}\right) =$$

$$\begin{aligned} &= \dim (F\mathcal{Q}^k + \text{ im } G) + \dim H[\mathcal{Q}^k \cap F^{-1}[\text{ im } G]] \\ &= \dim X + \dim H[\mathcal{Q}^k \cap F^{-1}[\text{ im } G]]. \end{aligned}$$

Consequently,

$$\dim H[\mathcal{Q}^k \cap F^{-1}[\text{ im } G]] = \dim \left(\begin{bmatrix} F \\ H \end{bmatrix} \mathcal{Q}^k + \text{ im } \begin{bmatrix} G \\ 0 \end{bmatrix}\right) - \dim X$$

so that

$$\pi_k = \dim \left(\begin{bmatrix} F \\ H \end{bmatrix} \mathcal{Q}^{k-1} + \text{ im } \begin{bmatrix} G \\ 0 \end{bmatrix}\right) - \dim \left(\begin{bmatrix} F \\ H \end{bmatrix} \mathcal{Q}^k + \text{ im } \begin{bmatrix} G \\ 0 \end{bmatrix}\right).$$

Observing that the iteration (6.5) coincides with the iteration (6.8) for the pencil (6.11), we conclude from Theorem 6.2 that the right-hand side of the above equation coincides with the number of left Kronecker indices of (6.11) that are equal to k.

Example 6.4 The mass-spring-damper system of Example 1.3 has one observability index, which is 2, as can be seen from the AR representation (1.5). To illustrate Corollary 6.3, we will calculate the observability index from the P representation (1.6). Note that condition (6.12) is certainly satisfied because of the minimality of the representation. As a result, we can apply Corollary 6.3 and calculate the observability index as the left Kronecker index of the matrix

$$\begin{bmatrix} sG - F \\ H \end{bmatrix} = \begin{bmatrix} s+c & -1 & 0 \\ k & s & -1 \\ 1 & 0 & 0 \\ 0 & 0 & 1 \end{bmatrix}. \tag{6.14}$$

The vector

$$g(s) = \begin{bmatrix} -s \\ 1 \\ -s^2 - cs - k \\ 1 \end{bmatrix}$$

is a minimal polynomial basis of the left null space of (6.14). As a result, the left Kronecker index equals 2, as it should.

6.1.2 Results for DP representations

In this section we express the nonzero observability indices in terms of the matrices K, L and M of a DP representation

$$\sigma K \xi = L\xi + Mw.$$

As before, K and L are mappings from X to Z, and M is a mapping from W to Z. The results in this subsection are obtained as corollaries of results in the previous subsection by rewriting (K, L, M) as a P representation (F, G, H) defined by

$$G := [K \quad 0], \quad F := [L \quad M], \quad H := [0 \quad I]. \tag{6.15}$$

Theorem 6.5 Let $\Sigma = (\mathbf{T}, W, \mathcal{B}) \in \mathcal{L}^q$ and let $\mathcal{B}$ be given by a DP representation (K, L, M). Let X^* be the limit space of the iteration

$$X^0 = X, \quad X^{m+1} = L^{-1}[KX^m + \text{ im } M] \tag{6.16}$$

and let $Q^k \subset X$ be given by the iteration

$$Q^0 = X, \quad Q^{m+1} = K^{-1}LQ^m. \tag{6.17}$$

Denote the number of observability indices of Σ that are equal to k by π_k. Then we have, for all $k \geq 1$,

$$\pi_k = \dim M^{-1}[LQ^{k-1} + KX^*] - \dim M^{-1}[LQ^k + KX^*].$$

Proof Write (K, L, M) as the P representation (F, G, H) with F, G and H defined as in (6.15). Let $\mathcal{Z}^*$ and $\mathcal{Q}^k$ be defined from F, G and H as in Theorem 6.1. Clearly we have, for all $k \geq 0$,

$$\mathcal{Q}^k = Q^k \oplus W.$$

Also, it is easily verified that

$$\mathcal{Z}^* = \{(x, w) \in X \oplus W \mid Lx + Mw \in KX^*\}. \tag{6.18}$$

The desired result now follows from Theorem 6.1. ◇

Through the connection (6.15) the nonredundancy condition that corresponds to (6.12) in the previous subsection is

$$[sK - L \quad M] \text{ has full row rank and has no zeros at infinity.} \tag{6.19}$$

The next theorem is therefore an immediate result of Corollary 6.3.

Theorem 6.6 Let $\Sigma = (\mathbf{T}, W, \mathcal{B}) \in \mathcal{L}^q$ and let $\mathcal{B}$ be given by a DP representation (K, L, M) for which (6.19) holds. Then the nonzero observability indices of Σ coincide with the left Kronecker indices of the pencil $sK - L$.

Proof Write (K, L, M) as the P representation (F, G, H) with F, G and H defined as in (6.15). It follows from Corollary 6.3 that the nonzero observability indices of Σ coincide with the left Kronecker indices of the pencil

$$\begin{bmatrix} sK - L & -M \\ 0 & I \end{bmatrix}. \tag{6.20}$$

Clearly, the left Kronecker indices of (6.20) coincide with the left Kronecker indices of the pencil

$$\begin{bmatrix} sK - L & 0 \\ 0 & I \end{bmatrix} \tag{6.21}$$

that is obtained from left multiplication of (6.20) by the nonsingular matrix

$$\begin{bmatrix} I & M \\ 0 & I \end{bmatrix}.$$

The left Kronecker indices of (6.21) evidently coincide with the left Kronecker indices of the pencil $sK - L$, and this proves the theorem.

Example 6.7 Let us consider the mass-spring-damper system of Example 1.3 again. To illustrate Theorem 6.6, we derive the observability index from the DP representation (1.7) (note that condition (6.19) is certainly satisfied because of the minimality of the representation). For this purpose, we calculate the left Kronecker index of the matrix

$$sK - L = \begin{bmatrix} s+c & -1 \\ k & s \\ -1 & 0 \end{bmatrix}. \tag{6.22}$$

The vector

$$g(s) = \begin{bmatrix} s \\ 1 \\ s^2 + cs + k \end{bmatrix}$$

is a minimal polynomial basis of the left null space of (6.22). As a result, the left Kronecker index equals 2, as it should.

6.1.3 Results for D representations

In this section we express the observability indices in terms of the matrices E, A, B, C and D of a D representation

$$\begin{aligned} \sigma E\xi &= A\xi + Bu \\ y &= C\xi + Du. \end{aligned}$$

As before, E and A are mappings from X_d to X_e, B is a mapping from U to X_e, C is a mapping from X_d to Y, and D is a mapping from U to Y. In our setting there are two obvious ways to obtain results for D representations. The first method proceeds by rewriting the D representation (E, A, B, C, D) as the P representation (F, G, H) with

$$G = [E \quad 0], \quad F = [A \quad B], \quad H = \begin{bmatrix} C & D \\ 0 & I \end{bmatrix}. \tag{6.23}$$

Results are then obtained as corollaries of results in Subsection 6.1.1. The second method proceeds by rewriting (E, A, B, C, D) as the DP representation (K, L, M) with

$$K = \begin{bmatrix} E \\ 0 \end{bmatrix}, \quad L = \begin{bmatrix} A \\ -C \end{bmatrix}, \quad M = \begin{bmatrix} 0 & B \\ I & -D \end{bmatrix} \tag{6.24}$$

and leads to corollaries of results in Subsection 6.1.2. The next theorem is then immediate from Theorem 6.5.

Theorem 6.8 Let $\Sigma = (\mathbf{T}, Y \oplus U, \mathcal{B}) \in \mathcal{L}^{p+m}$ and let $\mathcal{B}$ be given by a D representation (E, A, B, C, D). Let X^* be the limit space of the iteration

$$X^0 = X_d, \quad X^{m+1} = A^{-1}[EX^m + \text{ im } B] \tag{6.25}$$

and let $Q^k \subset X_d$ be given by the iteration

$$Q^0 = X_d, \quad Q^{m+1} = E^{-1}A[Q^m \cap \text{ ker } C]. \tag{6.26}$$

Denote the number of observability indices of Σ that are equal to k by π_k. Then we have, for all $k \geq 1$,

$$\pi_k = \ell_{k-1} - \ell_k,$$

where ℓ_k is given by

$$\ell_k = \dim \{(y,u) \in Y \oplus U \mid \exists x \in Q^k \text{ with } Ax + Bu \in EX^* \\ \text{and } y = Cx + Du\}.$$

◇

Through the connection (6.23) the nonredundancy condition that corresponds to (6.12) (or, equivalently, to (6.19)) is

$$[sE - A \quad B] \text{ has full row rank and has no zeros at infinity.} \tag{6.27}$$

The next theorem is therefore immediate from Corollary 6.3.

Theorem 6.9 Let $\Sigma = (\mathbf{T}, Y \oplus U, \mathcal{B}) \in \mathcal{L}^{p+m}$ and let $\mathcal{B}$ be given by a D representation (E, A, B, C, D) for which (6.27) holds. Then the nonzero observability indices of Σ coincide with the left Kronecker indices of the pencil

$$\begin{bmatrix} sE - A \\ C \end{bmatrix}.$$

Example 6.10 Consider the two interconnected electrical circuits of Example 1.6. To illustrate Theorem 6.9, we derive the observability indices from the D representation (1.15) (note that condition (6.27) is certainly satisfied because of the minimality of the representation). Let us therefore calculate the left Kronecker indices of the matrix

$$\begin{bmatrix} sE - A \\ C \end{bmatrix} = \begin{bmatrix} s \\ 0 \\ 1 \\ 1 \end{bmatrix}. \tag{6.28}$$

A minimal polynomial basis matrix for the left null space of (6.28) is given by

$$\begin{bmatrix} 1 & 1 & 0 \\ 0 & 0 & 1 \\ -s & 0 & 0 \\ 0 & -s & 0 \end{bmatrix}.$$

As a result, there are three left Kronecker indices. Their values are 1, 1 and 0.

Example 6.11 Consider the electrical circuit of Example 1.5, which is given by the D representation (1.11). Although (1.11) is not minimal, it does satisfy condition (6.27) (use Lemma 2.37 to check this). Consequently, we can apply Theorem 6.9 and calculate the observability index as the left Kronecker index of the matrix

$$\begin{bmatrix} sE - A \\ C \end{bmatrix} = \begin{bmatrix} cr_2 s + 1 & -r_2 \\ -1 & -r_1 \\ 1 & 0 \end{bmatrix}. \tag{6.29}$$

The vector

$$\begin{bmatrix} r_1 \\ -r_2 \\ -cr_1 r_2 s - r_1 - r_2 \end{bmatrix}$$

is a minimal polynomial basis of the left null space of (6.29) so that the left Kronecker index equals 1. ◇

For systems in standard state space form, i.e., given by a D representation with $E = I$, the nonredundancy condition (6.27) is automatically fulfilled. It therefore follows without any further assumptions that the observability indices of the system are given by the left Kronecker indices of the pencil

$$\begin{bmatrix} sI - A \\ C \end{bmatrix}. \tag{6.30}$$

For these systems, the observability indices, as defined in this book, therefore coincide with the classical notion of observability indices as given by [71, 72] (cf. Remark 2.32). Note that the matrix B is completely irrelevant in the context of standard state space representations, whereas the above example shows that this is not necessarily the case for descriptor representations.

6.2 Controllability indices

In this section we consider the controllability indices of a system Σ in $\mathcal{L}^q$. These indices were defined in Definition 3.11; they are invariant under weak external equivalence. Our aim is to derive expressions for the controllability indices in terms of the first-order representations that we considered previously. In the previous subsection we showed that the observability indices of a system in standard state space form coincide with the classical notions and that this result holds without any further assumptions. With respect to controllability indices, one might conjecture that a similar result holds. More specifically, the conjecture would be that the controllability indices of a system in standard state space form are given by the right Kronecker indices of the matrix pencil $[sI - A \quad B]$. We will see in this section that this is indeed the case; however, in contrast with the previous section, it will be necessary to impose certain restrictions on the representation for the result to hold. It is indeed easily seen that the result cannot be valid without any further restrictions: for $C = 0$ the controllability indices of the system are trivial, whereas the right Kronecker indices of $[sI - A \quad B]$ need not be trivial.

The above is explained by the fact that we are dealing with a set-up in which controllability and observability indices do not play dual roles. However, for systems in the restricted class of controllable systems $\mathcal{L}_c^q$, the two sets of indices *do* play dual roles (see Theorem 3.22). The development below is based on this duality. Before presenting expressions for controllability indices in terms of the pencil and the dual pencil form, we first state a lemma, according to which observability indices of controllable systems are expressed in terms of pencil matrices. The lemma is a key result in the development in the next subsection.

Lemma 6.12 Let $\Sigma = (\mathbf{T}, W, \mathcal{B}) \in \mathcal{L}_c^q$ with $\mathcal{B}$ given by the MA representation

$$w = Q(\sigma)\xi. \tag{6.31}$$

Let (F, G, H) be a P representation that is weakly externally equivalent to (6.31). Let $\mathcal{Z}^*$ be the limit space of the iteration (6.3) and let $\bar{\mathcal{Q}}^k$ be given by the iteration

$$\bar{\mathcal{Q}}^0 = \mathcal{T}^*, \quad \bar{\mathcal{Q}}^{m+1} = G^{-1}F[\bar{\mathcal{Q}}^m \cap \ker H]$$

where $\mathcal{T}^*$ is the limit space of the iteration

$$\mathcal{T}^0 = \{0\}, \quad \mathcal{T}^{m+1} = G^{-1}F\mathcal{T}^m.$$

Denote the number of observability indices of Σ that are equal to k by π_k. Then we have, for all $k \geq 1$,

$$\pi_k = \dim H[\bar{\mathcal{Q}}^{k-1} \cap \mathcal{Z}^*] - \dim H[\bar{\mathcal{Q}}^k \cap \mathcal{Z}^*].$$

Proof According to Lemma 3.16, the RC space $\mathcal{C}$ is given as

$$\mathcal{C} = \text{ im } M(\lambda),$$

whereas the RB space $\mathcal{W}_0$ is given as

$$\mathcal{W}_0 = \{w(\lambda) \in \lambda^{-1}W_\infty(\lambda) \mid \exists\tilde{\xi}(\lambda) \text{ with } w(\lambda) - M(\lambda)\tilde{\xi}(\lambda) \text{ is polynomial}\}.$$

It follows that

$$\mathcal{W}_0 = \{w(\lambda) \in \lambda^{-1}W_\infty(\lambda) \mid \exists\xi(\lambda) \in \mathcal{C} \text{ with } w(\lambda) = \pi_-\xi(\lambda)\}.$$

Since the P representation (F, G, H) is weakly externally equivalent to representation (6.31), we have $\mathcal{C} = H[\ker(\lambda G - F)]$ so that W_k^0 as defined in (6.1) is given by

$$W_k^0 = \pi_1\{w(\lambda) \in \lambda^{-1}W_\infty(\lambda) \mid \exists \xi(\lambda) \in Z(\lambda) \text{ with } \lambda^{-k}w(\lambda) = \pi_-(H\xi(\lambda)) \\ \text{and } (\lambda G - F)\xi(\lambda) = 0\}. \tag{6.32}$$

Because of (6.2), it is obviously sufficient to prove that, for all $k \geq 0$,

$$W_k^0 = H[\bar{\mathcal{Q}}^k \cap \mathcal{Z}^*].$$

For this, let $k \geq 0$ and let $w \in W_k^0$; we will first prove that $W_k^0 \subset H[\bar{\mathcal{Q}}^k \cap \mathcal{Z}^*]$. Because of (6.32) and the fact that $w \in W_k^0$, there exist $w(\lambda) \in \lambda^{-1}W_\infty(\lambda)$ and $\xi(\lambda) \in Z(\lambda)$ such that

$$(\lambda G - F)\xi(\lambda) = 0 \tag{6.33}$$

and

$$w(\lambda) = \lambda^k \pi_-(H\xi(\lambda)). \tag{6.34}$$

Let us write the Laurent expansion of $\xi(\lambda)$ as

$$\xi(\lambda) = \xi_{-\ell}\lambda^\ell + \xi_{-\ell+1}\lambda^{\ell-1} + \cdots + \xi_0 + \xi_1\lambda^{-1}\cdots.$$

It follows from (6.33) that $G\xi_{i+1} = F\xi_i$ for all $i \geq -\ell$ so that $\xi_i \in \mathcal{Z}^*$ for all $i \geq -\ell$. Also, $G\xi_{-\ell} = 0$ so that $\xi_i \in \mathcal{T}^{\ell+i+1}$ for all $i \geq -\ell$. In particular $\xi_1 \in \mathcal{T}^{\ell+2}$, so certainly $\xi_1 \in \mathcal{T}^*$. From (6.34) it follows that $w = H\xi_{k+1}$ and that $H\xi_i = 0$ for $1 \leq i \leq k$. We conclude that $\xi_{k+1} \in \bar{\mathcal{Q}}^k \cap \mathcal{Z}^*$ so that $w \in H[\bar{\mathcal{Q}}^k \cap \mathcal{Z}^*]$.

Conversely, let $w = Hz$ with $z \in \bar{\mathcal{Q}}^k \cap \mathcal{Z}^*$. Since $z \in \mathcal{Z}^*$, it follows from Lemma 2.29 (use Remark 2.30) that there exists a strictly proper rational function $\tilde{\xi}(\lambda)$ such that $(\lambda G - F)\tilde{\xi}(\lambda) = Fz$. Because of the fact that $z \in \bar{\mathcal{Q}}^k$, we can find $\xi_i \in Z$ for $1 \leq i \leq k$ such that

$$Gz = F\xi_k,\ G\xi_k = F\xi_{k-1}, \ldots, G\xi_2 = F\xi_1$$

while

$$H\xi_k = H\xi_{k-1} = \cdots = H\xi_1 = 0 \text{ and } \xi_1 \in \mathcal{T}^*.$$

Since $\xi_1 \in \mathcal{T}^*$, there exist $\xi_0, \xi_{-1}, \ldots, \xi_{-\ell}$ such that $G\xi_{-\ell} = 0$ and

$$G\xi_1 = F\xi_0,\ G\xi_0 = F\xi_{-1}, \ldots, G\xi_{-\ell+1} = F\xi_\ell.$$

Now define

$$\xi(\lambda) := \xi_{-\ell}\lambda^\ell + \cdots + \xi_0 + \xi_1\lambda^{-1} + \cdots + \xi_k\lambda^{-k} + (z + \tilde{\xi}(\lambda))\lambda^{-k-1}.$$

Then $(\lambda G - F)\tilde{\xi}(\lambda) = 0$ and $w(\lambda) := \lambda^k\pi_-(H\xi(\lambda))$ is strictly proper with $\pi_1 w(\lambda) = Hz = w$. It now follows from (6.32) that $w \in W_k^0$. We conclude that $W_k^0 = H[\bar{\mathcal{Q}}^k \cap \mathcal{Z}^*]$, which proves the theorem. ◇

In the following subsections we will derive expressions for the nonzero controllability indices in terms of DP representations, P representations and D representations.

6.2.1 Results for DP representations and P representations

In this section we consider the controllability indices of systems that are either given in dual pencil form or in pencil form. Our strategy will be based on the duality that exists under weak external equivalence between DP representations and P representations (Lemma 3.26). By this duality, the next theorem is a corollary of Lemma 6.12; the theorem gives a geometric formula for the nonzero controllability indices in terms of the matrices K, L and M of a DP representation

$$\sigma K\boldsymbol{\xi} = L\boldsymbol{\xi} + M\boldsymbol{w}.$$

As before, K and L are mappings from X to Z, and M is a mapping from W to Z.

Theorem 6.13 Let $\Sigma = (\mathbf{T}, W, \mathcal{B}) \in \mathcal{L}^q$ and let $\mathcal{B}$ be given by a DP representation (K, L, M). Let T^* be the limit space of the iteration

$$T^0 = \{0\}, \quad T^{m+1} = K^{-1}LT^m,$$

and let $\bar{R}^k \subset X$ be given by the iteration

$$\bar{R}^0 = V^*, \quad \bar{R}^{m+1} = L^{-1}[K\bar{R}^m + \text{ im } M],$$

where V^* is the limit space of the iteration

$$V^0 = X, \quad V^{m+1} = L^{-1}KV^m. \tag{6.35}$$

Denote the number of controllability indices of Σ that are equal to k by π_k. Then we have, for all $k \geq 1$,

$$\pi_k = \dim M^{-1}[K\bar{R}^k + LT^*] - \dim M^{-1}[K\bar{R}^{k-1} + LT^*].$$

Proof Let $\mathcal{B}$ be given by the AR representation

$$R(\sigma)\boldsymbol{w} = 0.$$

Let $\tilde{\Sigma} = (\mathbf{T}, W, \tilde{\mathcal{B}})$ be the system in $\mathcal{L}^q_c$ whose behavior $\tilde{\mathcal{B}}$ is given by the MA representation

$$\boldsymbol{w} = R^T(\sigma)\boldsymbol{\xi}. \tag{6.36}$$

It follows from Lemma 3.23 that the controllability indices of Σ coincide with the observability indices of $\tilde{\Sigma}$. Moreover, we conclude from Lemma 3.26 that the P representation (L^T, K^T, M^T) is weakly externally equivalent to the MA representation (6.36). By the previous lemma we therefore have a geometric formula for the controllability indices of Σ in terms of the matrices K^T, L^T and M^T. The desired result then follows from elementary results from linear algebra. ◇

An analogous formula for P representations can now be obtained by rewriting a P representation (F, G, H) in dual pencil form (K, L, M) with

$$K = \begin{bmatrix} G \\ 0 \end{bmatrix}, \quad L = \begin{bmatrix} F \\ H \end{bmatrix}, \quad M = \begin{bmatrix} 0 \\ -I \end{bmatrix}. \tag{6.37}$$

Before presenting the next theorem, let us introduce the iteration

$$\mathcal{N}^0 = \{0\}, \quad \mathcal{N}^{m+1} = G^{-1}F[\mathcal{N}^m \cap \ker H]. \tag{6.38}$$

Note that this iteration coincides with iteration (2.29) applied to the pencil

$$\begin{bmatrix} sG - F \\ H \end{bmatrix}. \tag{6.39}$$

The limit space $\mathcal{N}^*$ is related to redundancy in the representation (F, G, H): when $\ker G \cap \ker H = \{0\}$, then $\mathcal{N}^* = \ker G$. More precisely, Lemma 2.36 yields the following implication:

$$F^{-1}[\operatorname{im} G] \cap \ker G \cap \ker H = \{0\} \implies \mathcal{N}^* = \ker G. \tag{6.40}$$

Theorem 6.14 Let $\Sigma = (\mathbf{T}, W, \mathcal{B}) \in \mathcal{L}^q$ and let $\mathcal{B}$ be given by a P representation (F, G, H). Let $\mathcal{N}^*$ be defined as the limit space of the iteration (6.38) and let $\mathcal{V}^*$ be the limit space of the iteration

$$\mathcal{V}^0 = Z, \quad \mathcal{V}^{m+1} = (F^{-1}G\mathcal{V}^m) \cap \ker H. \tag{6.41}$$

Let $\bar{\mathcal{R}}^k$ be given by the iteration

$$\bar{\mathcal{R}}^0 = \mathcal{V}^*, \quad \bar{\mathcal{R}}^{m+1} = F^{-1}G\bar{\mathcal{R}}^m. \tag{6.42}$$

Denote the number of controllability indices of Σ that are equal to k by π_k. Then we have, for all $k \geq 1$,

$$\pi_k = \dim H[\bar{\mathcal{R}}^{k+1} \cap \mathcal{N}^*] - \dim H[\bar{\mathcal{R}}^k \cap \mathcal{N}^*]. \tag{6.43}$$

Proof Write (F, G, H) as the DP representation (K, L, M) with K, L and M defined as in (6.37). Let T^*, V^* and $\bar{R}^k$ be defined from K, L and M as in Theorem 6.13. It is easily verified that $V^* = \mathcal{V}^*$ and that $T^* = \mathcal{N}^*$. Also, for all $k \geq 0$,

$$\bar{R}^k = \bar{\mathcal{R}}^k.$$

We conclude from Theorem 6.13 that

$$\pi_k = \dim H[F^{-1}G\bar{\mathcal{R}}^k \cap \mathcal{N}^*] - \dim H[F^{-1}G\bar{\mathcal{R}}^{k-1} \cap \mathcal{N}^*],$$

which coincides with (6.43). ◇

Note that Theorem 6.13 and Theorem 6.14 in fact present recursive formulas for the nonzero minimal indices of rational vector spaces of the form $M^{-1}[\operatorname{im}(sK - L)]$ and $H[\ker(sG - F)]$, respectively (use Lemma 3.16).

Expressions in terms of matrix pencils can be obtained by imposing certain nonredundancy conditions as in Section 6.1. Roughly speaking, the nonredundancy conditions for P representations should rule out the effect of *both* $\mathcal{N}^*$ and $\mathcal{V}^*$. Thus the situation is different from the situation in Subsection 6.1.1, where we only had to take care of the effect of one space, namely $\mathcal{Z}^*$.

Corollary 6.15 Let $\Sigma = (\mathbf{T}, W, \mathcal{B}) \in \mathcal{L}^q$ and let $\mathcal{B}$ be given by a P representation (F, G, H) for which

$$\begin{bmatrix} sG - F \\ H \end{bmatrix} \text{ has no zeros in } \mathbb{C}^e. \tag{6.44}$$

Then the nonzero controllability indices of Σ coincide with the right Kronecker indices of the pencil $sG - F$.

Proof Let $\mathcal{N}^*$, $\mathcal{V}^*$, $\bar{\mathcal{R}}^k$ and π_k be defined as in Theorem 6.14. By Lemma 2.36 it follows from (6.44) that $\mathcal{N}^* = \ker G$ and that

$$F^{-1}[\operatorname{im} G] \cap \ker G \cap \ker H = \{0\}. \tag{6.45}$$

According to Theorem 6.14, the value of π_k $(k \geq 1)$ is then given by

$$\pi_k = \dim H[\bar{\mathcal{R}}^{k+1} \cap \ker G] - \dim H[\bar{\mathcal{R}}^k \cap \ker G].$$

Because of (6.45) and the fact that $\bar{\mathcal{R}}^k \subset F^{-1}[\operatorname{im} G]$ for all $k \geq 0$, we have that $\dim H[\bar{\mathcal{R}}^k \cap \ker G] = \dim (\bar{\mathcal{R}}^k \cap \ker G)$ for all $k \geq 0$ so that

$$\pi_k = \dim (\bar{\mathcal{R}}^{k+1} \cap \ker G) - \dim (\bar{\mathcal{R}}^k \cap \ker G).$$

The iteration (6.41) of which $\mathcal{V}^*$ is the limit space coincides with the iteration (2.24) applied to the pencil (6.39). Consequently, it follows from (6.44) that $\mathcal{V}^* = \{0\}$ (Theorem 2.31) so that the iteration (6.42) coincides with the iteration (6.9) applied to the pencil $sG - F$. The desired result is now immediate from Theorem 6.2.

Example 6.16 The controllability index of the mass-spring-damper system of Example 1.3 can be calculated from the P representation (1.6). It can be easily checked that condition (6.44) is satisfied so that we can apply Corollary 6.15: the controllability index equals the right Kronecker index of the matrix

$$sG - F = \begin{bmatrix} s + c & -1 & 0 \\ k & s & -1 \end{bmatrix}. \tag{6.46}$$

The vector

$$g(s) = \begin{bmatrix} 1 \\ s + c \\ s^2 + cs + k \end{bmatrix}$$

is a minimal polynomial basis of the kernel of (6.46) so that the right Kronecker index equals 2.

Example 6.17 Consider the simplified model of a satellite with flexible appendages of Example 5.14. A minimal P representation of this system is given by (5.23). This representation does not satisfy condition (6.44). Consequently, we can not apply Corollary 6.15 to this representation. From the MA representation (5.24) of the controllable part of the behavior we know that the system has one controllability index, which has value 4 (use Theorem 3.22 and the fact that controllability indices are invariant under weak external equivalence). To illustrate Corollary 6.15, we calculate the controllability index from the P representation (5.25) of the controllable subbehavior, which satisfies condition (6.44). Applying Corollary 6.15 to (5.25), we find that the matrix

$$sG - F = \begin{bmatrix} s & -1 & 0 & 0 & 0 \\ 0 & s & -1 & 0 & 0 \\ 0 & 0 & s & -1 & 0 \\ 0 & 0 & 0 & s & -1 \end{bmatrix} \tag{6.47}$$

has one right Kronecker index, which has value 4: the vector

$$\begin{bmatrix} 1 \\ s \\ s^2 \\ s^3 \\ s^4 \end{bmatrix}$$

is a minimal polynomial basis of the kernel of (6.47). We conclude that the system has indeed controllability index 4. ◇

An analogous result for DP representations is obtained by rewriting the DP representation (K, L, M) in pencil form as (6.15). The nonredundancy condition that corresponds to (6.44) is

$$sK - L \text{ has no zeros in } \mathbb{C}^e. \tag{6.48}$$

The next theorem is an immediate result of Corollary 6.15 and will therefore be given without proof.

Theorem 6.18 Let $\Sigma = (\mathbf{T}, W, \mathcal{B}) \in \mathcal{L}^q$ and let $\mathcal{B}$ be given by a DP representation (K, L, M) for which (6.48) holds. Then the nonzero controllability indices of Σ coincide with the right Kronecker indices of the pencil $[sK - L \quad M]$.

Example 6.19 The controllability index of the mass-spring-damper system of Example 1.3 can be calculated from the DP representation (1.7). It can be easily checked that condition (6.48) is satisfied so that we can apply

Theorem 6.18: the controllability index equals the right Kronecker index of the matrix

$$\begin{bmatrix} sK - L & M \end{bmatrix} = \begin{bmatrix} s+c & -1 & 0 & 0 \\ k & s & 0 & 1 \\ -1 & 0 & -1 & 0 \end{bmatrix}. \tag{6.49}$$

The vector

$$\begin{bmatrix} 1 \\ s+c \\ -1 \\ -s^2 - cs - k \end{bmatrix}$$

is a minimal polynomial basis of the kernel of (6.49) so that the right Kronecker index equals 2.

6.2.2 Results for D representations

In this section we express the controllability indices in terms of the matrices E, A, B, C and D of a D representation

$$\begin{aligned} \sigma E\xi &= A\xi + Bu \\ y &= C\xi + Du. \end{aligned}$$

As before, E and A are mappings from X_d to X_e, B is a mapping from U to X_e, C is a mapping from X_d to Y, and D is a mapping from U to Y. Proceeding as in Subsection 6.1.3, we can either rewrite the D representation (E, A, B, C, D) in dual pencil form (6.24) and obtain a geometric formula from Theorem 6.13 or rewrite (E, A, B, C, D) in pencil form (6.23) and obtain a geometric formula from Theorem 6.14. The proof of the next theorem is based on Theorem 6.14.

Theorem 6.20 Let $\Sigma = (\mathbf{T}, Y \oplus U, \mathcal{B}) \in \mathcal{L}^{p+m}$ and let $\mathcal{B}$ be given by a D representation (E, A, B, C, D). Let N^* be the limit space of the iteration

$$N^0 = \{0\}, \quad N^{m+1} = E^{-1}A[N^m \cap \ker C], \tag{6.50}$$

and let $\bar{R}^k \subset X_d$ be given by the iteration

$$\bar{R}^0 = V^*, \quad \bar{R}^{m+1} = A^{-1}[E\bar{R}^m + \operatorname{im} B], \tag{6.51}$$

where V^* is the limit space of the iteration

$$V^0 = X_d, \quad V^{m+1} = (A^{-1}EV^m) \cap \ker C. \tag{6.52}$$

Denote the number of controllability indices of Σ that are equal to k by π_k. Then we have, for all $k \geq 1$,

$$\pi_k = \ell_k - \ell_{k-1},$$

where ℓ_k is given by

$$\ell_k = \dim \{(y, u) \in Y \oplus U \mid \exists x \in N^* \text{ with } Ax + Bu \in E\bar{R}^k \text{ and } y = Cx + Du\}.$$

Proof Write (E, A, B, C, D) as the P representation (F, G, H) with F, G and H defined as in (6.23). Let $\mathcal{N}^*$, $\mathcal{V}^*$ and $\bar{\mathcal{R}}^k$ be defined from F, G and H as in Theorem 6.14. It is easily verified that $\mathcal{N}^* = N^* \oplus U$ and that $\mathcal{V}^* = V^* \oplus \{0\}$. Also, for all $k \geq 0$,

$$\bar{\mathcal{R}}^{k+1} = \{(x, u) \in X_d \oplus U \mid Ax + Bu \in E\bar{R}^k\}.$$

The desired result now follows from Theorem 6.14. ◇

Through the connection (6.24) the nonredundancy condition that corresponds to (6.48) (or, equivalently, to (6.44)) is

$$\begin{bmatrix} sE - A \\ C \end{bmatrix} \text{ has no zeros in } \mathbb{C}^e. \tag{6.53}$$

The next theorem is immediate from Theorem 6.18.

Theorem 6.21 Let $\Sigma = (\mathbf{T}, Y \oplus U, \mathcal{B}) \in \mathcal{L}^{p+m}$ and let $\mathcal{B}$ be given by a D representation (E, A, B, C, D) for which (6.53) holds. Then the nonzero controllability indices of Σ coincide with the right Kronecker indices of the pencil $[sE - A \quad B]$. ◇

The above theorem generalizes a result in [28], where the authors essentially consider the controllability indices of an I-O system that is given by a DZ representation (E, A, B, C) with $C = I$. The condition (6.53) is then automatically fulfilled. Controllability indices are used in [20] to derive a certain canonical form of (E, A, B, I).

Example 6.22 The controllability index of the two interconnected electrical circuits of Example 1.6 can be calculated from the D representation (1.15). It can be easily checked that condition (6.53) is satisfied so that we can apply Theorem 6.21: the controllability index equals the right Kronecker index of the matrix

$$[sE - A \quad B] = \begin{bmatrix} s & 0 & 1 \\ 0 & 1 & -c_1 \end{bmatrix}. \tag{6.54}$$

The vector

$$\begin{bmatrix} 1 \\ -c_1 s \\ -s \end{bmatrix}$$

is a minimal polynomial basis of the kernel of (6.54) so that the right Kronecker index equals 1.

Example 6.23 Consider the electrical circuit of Example 1.5, which is given by the (nonminimal) D representation (1.11). The representation satisfies condition (6.53) (use Lemma 2.36). Consequently, we may apply

Theorem 6.21 to obtain the controllability index as the right Kronecker index of the matrix

$$[sE - A \quad B] = \begin{bmatrix} cr_2 s + 1 & -r_2 & 0 \\ -1 & -r_1 & -1 \end{bmatrix}. \tag{6.55}$$

The vector

$$\begin{bmatrix} r_2 \\ cr_2 s + 1 \\ -cr_1 r_2 s - r_1 - r_2 \end{bmatrix}$$

is a minimal polynomial basis of the kernel of (6.55) so that the right Kronecker index equals 1. As a result, the system has one controllability index, which has value 1. In case that $r_1 = 0$, the system is alternatively described by the MA representation

$$\begin{bmatrix} \boldsymbol{y} \\ \boldsymbol{u} \end{bmatrix} = \begin{bmatrix} 1 \\ 1 \end{bmatrix} \boldsymbol{\ell}$$

so that there is one controllability index, which has value 0. Note, however, that the right Kronecker index of the matrix

$$[sE - A \quad B] = \begin{bmatrix} cr_2 s + 1 & -r_2 & 0 \\ -1 & 0 & -1 \end{bmatrix}$$

equals 1. This seems to contradict Theorem 6.21; however, the theorem cannot be applied since condition (6.53) is not fulfilled. ◇

For systems in standard state space form, i.e., given by a D representation with $E = I$, it follows that the controllability indices are given by the right Kronecker indices of the pencil $[sI - A \quad B]$ if the representation is observable. Thus for systems in observable standard form, the controllability indices, as defined in this book, coincide with the classical notion of controllability indices as given by [71, 72] (cf. Remark 2.32).

6.3 The input-output structure

Our starting point in this section is as in Section 3.5: we consider the situation in which the external variables are split into two parts. As before, we use the notation $W = Y \oplus U$, where $Y = \mathbb{R}^p$ and $U = \mathbb{R}^m$ and we do not adhere any special meaning to this decomposition. The question arises: when is a system $\Sigma \in \mathcal{L}^{p+m}$ an I-O system as defined by Definition 3.27? First, there is the requirement that there should exist at least one transfer function for the system. Following the terminology of [67, 68], we will call $\boldsymbol{u}$ "free" if this requirement is satisfied. Second, there is the requirement that $\boldsymbol{y}$-trajectories should be determined by $\boldsymbol{u}$-trajectories, up to initial conditions. This requirement takes care of the uniqueness; if it is satisfied we will say, again in accordance with [67, 68], that $\boldsymbol{y}$ "processes" $\boldsymbol{u}$. Precise definitions of "processing" and "free" will be given below. The definitions involve certain subspaces $W^k \subset W$ which are defined as follows:

Definition 6.24 Let $\Sigma = (\mathbf{T}, W = Y \oplus U, \mathcal{B}) \in \mathcal{L}^{p+m}$ and let $\mathcal{W}_*$ be defined as in Definition 3.10. For all $k \in \mathbb{Z}$, we define:

$$W^k = \pi_1\{\begin{bmatrix} y(\lambda) \\ u(\lambda) \end{bmatrix} \in \lambda^{-1}W_\infty(\lambda) \mid \begin{bmatrix} y(\lambda) \\ \lambda^{-k}u(\lambda) \end{bmatrix} \in \mathcal{W}_*\}.$$

◇

Note that for $k = 0$ the space W^0 is defined as in Definition 3.10 and that the spaces W^k are invariant under weak external equivalence. Our definition of W^k stems from [48], where subspaces of Y and U are used to determine the "zeros at infinity" of (proper) transfer functions. However, the subspaces in [48] are defined from the transfer function, whereas our definition of W^k is more general: it is also applicable for systems without a transfer function. We have the following lemma.

Lemma 6.25 The following statements hold for all $k \in \mathbb{Z}$:

(i) $\dim W^k = \dim W^0$

(ii) $\pi_Y W^k \subset \pi_Y W^{k-1}$

(iii) $\pi_U W^k \subset \pi_U W^{k+1}$

(iv) $\dim \pi_Y W^k + \dim \pi_U W^{k-1} = \dim W^0$.

Proof Equality (i) follows from (see the remarks following Definition 3.10)

$$\dim W^k = \dim \mathcal{W}_* = \dim W^0.$$

Next, the inclusions (ii) and (iii) are immediate from the definition of W^k while equality (iv) follows from

$$\begin{aligned}
\dim W^k &= \dim \pi_Y W^k + \dim \{\begin{bmatrix} y \\ u \end{bmatrix} \in W^k \mid y = 0\} \\
&= \dim \pi_Y W^k + \dim \pi_1\{u(\lambda) \in \lambda^{-1}U_\infty(\lambda) \mid \exists y(\lambda) \in \lambda^{-1}Y_\infty(\lambda) \\
&\qquad \text{such that } \lambda y(\lambda) \in \lambda^{-1}Y_\infty(\lambda) \text{ and } \begin{bmatrix} \lambda y(\lambda) \\ \lambda^{-k+1}u(\lambda) \end{bmatrix} \in \mathcal{W}_*\} \\
&= \dim \pi_Y W^k + \dim \pi_U W^{k-1}.
\end{aligned}$$

This completes the proof of the lemma. ◇

The above lemma shows that the sequence $\{\pi_Y W^k\}$ is nonincreasing and that the sequence $\{\pi_U W^k\}$ is nondecreasing. Since Y and U are finite-dimensional, both sequences must have limits both as $k \to \infty$ and as $k \to -\infty$. In obvious notation we shall denote the limit spaces by Y_+^*, Y_-^*, U_+^* and U_-^*, respectively. We then have the following situation:

$$Y_+^* \subset \cdots \subset \pi_Y W^{k+1} \subset \pi_Y W^k \subset \cdots \subset Y_-^*$$

$$U_-^* \subset \cdots \subset \pi_U W^k \subset \pi_U W^{k+1} \subset \cdots \subset U_+^*.$$

We are now ready to give precise definitions of the properties that were mentioned at the beginning of the section.

Definition 6.26 Let $\Sigma = (\mathbf{T}, W = Y \oplus U, \mathcal{B}) \in \mathcal{L}^{p+m}$. We define the following concepts:

(i) $\boldsymbol{y}$ *processes* $\boldsymbol{u}$ if $Y_+^* = \{0\}$

(ii) $\boldsymbol{u}$ is *free* if $U_+^* = U$.

◇

The next lemma shows that Σ is an I-O system, i.e., has a unique transfer function $T(s)$ if the above conditions (i) and (ii) hold at the same time.

Lemma 6.27 Let $\Sigma = (\mathbf{T}, Y \oplus U, \mathcal{B}) \in \mathcal{L}^{p+m}$ and let $\mathcal{B}$ be given by the AR representation

$$[R_1(\sigma) \quad R_2(\sigma)] \begin{bmatrix} \boldsymbol{y} \\ \boldsymbol{u} \end{bmatrix} = 0,$$

where $[R_1(s) \quad R_2(s)]$ is a matrix of size $r \times (p+m)$ of full row rank. Then the following hold:

(i) $\boldsymbol{y}$ processes $\boldsymbol{u}$ if and only if $R_1(s)$ has full column rank

(ii) $\boldsymbol{u}$ is free if and only if $R_1(s)$ has full row rank.

Moreover, Σ is an I-O system if and only if both (i) and (ii) hold; the transfer function $T(s)$ is then given by $T(s) = -R_1^{-1}(s)R_2(s)$.

Proof From the definition it follows that

$$W^k = \pi_1\{\begin{bmatrix} y(\lambda) \\ u(\lambda) \end{bmatrix} \in \lambda^{-1}W_\infty(\lambda) \mid \lambda^k R_1(\lambda)y(\lambda) + R_2(\lambda)u(\lambda) = 0\}. \quad (6.56)$$

From this we conclude that

$$\dim Y_+^* = \dim \ \ker R_1(s), \quad (6.57)$$

which yields (i). By Lemma 6.25 (iv) we have that

$$\begin{aligned}\dim U_+^* &= \dim W^0 - \dim Y_+^* \\ &= p+m-r-\dim\ \ker R_1(s) \\ &= m-r+\dim\ \operatorname{im} R_1(s),\end{aligned}$$

which yields (ii). Because $[R_1(s) \quad R_2(s)]$ has full row rank, it is clear that Σ is an I-O system if and only if $R_1(s)$ is invertible, which proves the last statement of the theorem.

Example 6.28 Consider the two mass-spring-damper systems of Example 1.2. Let $\boldsymbol{y} = [\boldsymbol{q}_1 \quad \boldsymbol{q}_2]^T$ and $\boldsymbol{u} = [\boldsymbol{F}_1 \quad \boldsymbol{F}_2]^T$ as usual. The interconnection of the two systems can be considered as a controller's action: the behavior of the controller is then described by

$$\begin{bmatrix} 0 & 0 & 1 & 1 \\ 1 & -1 & 0 & 0 \end{bmatrix} \begin{bmatrix} \boldsymbol{q}_1 \\ \boldsymbol{q}_2 \\ \boldsymbol{F}_1 \\ \boldsymbol{F}_2 \end{bmatrix} = \begin{bmatrix} 0 \\ 0 \end{bmatrix}. \tag{6.58}$$

For this system, neither $\boldsymbol{y}$ nor $\boldsymbol{u}$ is free. Furthermore, $\boldsymbol{y}$ does not process $\boldsymbol{u}$ and $\boldsymbol{u}$ does not process $\boldsymbol{y}$. This implies that the controller is not an I-O system. Actually, the system is not considered to be a controller in the classical sense: it has no transfer function. However, the behavioral approach enables one to deal with this system as a controller. For more details, the reader is referred to [36]. ◇

For duality purposes, we also give a description in terms of MA representations:

Lemma 6.29 Let $\Sigma = (\mathbf{T}, Y \oplus U, \mathcal{B}) \in \mathcal{L}^{p+m}$ and let $\mathcal{B}$ be given by the MA representation

$$\begin{bmatrix} \boldsymbol{y} \\ \boldsymbol{u} \end{bmatrix} = \begin{bmatrix} Q_1(\sigma) \\ Q_2(\sigma) \end{bmatrix} \boldsymbol{\xi},$$

where

$$\begin{bmatrix} Q_1(s) \\ Q_2(s) \end{bmatrix} \tag{6.59}$$

is a matrix of size $(p+m)\times r$ of full column rank. Then the following hold:

(i) $\boldsymbol{y}$ processes $\boldsymbol{u}$ if and only if $Q_2(s)$ has full column rank

(ii) $\boldsymbol{u}$ is free if and only if $Q_2(s)$ has full row rank.

Moreover, Σ is an I-O system if and only if it has a unique transfer function $T(s) = Q_1(s)Q_2^{-1}(s)$.

Proof From the definition it follows that

$$W^k = \pi_1\{\begin{bmatrix} y(\lambda) \\ u(\lambda) \end{bmatrix} \in \lambda^{-1}W_\infty(\lambda) \mid \begin{bmatrix} y(\lambda) \\ \lambda^{-k}u(\lambda) \end{bmatrix} \in \text{ im } \begin{bmatrix} Q_1(\lambda) \\ Q_2(\lambda) \end{bmatrix}\}. \quad (6.60)$$

From this we conclude that

$$\dim U_+^* = \text{rank } Q_2(s), \quad (6.61)$$

which yields (ii). By Lemma 6.25 (iv) we have that

$$\begin{aligned} \dim Y_+^* &= \dim W^0 - \dim U_+^* \\ &= r - \text{rank } Q_2(s) \\ &= \dim \ \ker Q_2(s), \end{aligned}$$

which yields (i). Because (6.59) has full column rank, it is clear that Σ is an I-O system if and only if $Q_2(s)$ is invertible, which proves the last statement of the theorem. ◇

The next lemma shows that information on the rank and pole/zero structure at infinity of $T(s)$ (Theorem 2.6) is contained in the sequences $\{\pi_Y W^k\}$ and $\{\pi_U W^k\}$. Poles/zeros at infinity of $T(s)$ are also called "transmission poles/zeros at infinity" in the literature.

Lemma 6.30 Let $\Sigma = (\mathbf{T}, Y \oplus U, \mathcal{B}) \in \mathcal{L}^{p+m}$ be an I-O system with transfer function $T(s)$. Denote the number of poles at infinity of $T(s)$ of order $\geq k$ ($\leq k$) by p_k (s_k) ($k \in \mathbb{Z}$). Then the following hold:

(i) $\dim \ker T(s) = \dim U_-^*$

(ii) $\text{rank } T(s) = \dim Y_-^*$

(iii) $p_k = \dim \pi_Y W^k$

(iv) $s_k = \dim \pi_U W^k - \dim U_-^*$.

Proof Let $R_1(s)$ and $R_2(s)$ be polynomial matrices of sizes $r \times r$ and $r \times m$, respectively, such that $T(s) = R_1^{-1}(s)R_2(s)$. Then $\dim \ker T(s) = \dim \ \ker R_2(s)$ and $\text{rank } T(s) = \text{rank } R_2(s)$. It follows as in the proof of Lemma 6.27 that

$$\dim U_-^* = \dim \ \ker R_2(s) \quad (6.62)$$

and

$$\dim Y_-^* = \text{rank } R_2(s).$$

This yields (i) and (ii). From (6.56) we conclude that the spaces W^k can be written as

$$W^k = \pi_1\{\begin{bmatrix} y(\lambda) \\ u(\lambda) \end{bmatrix} \in \lambda^{-1}W_\infty(\lambda) \mid y(\lambda) = \lambda^{-k}T(\lambda)u(\lambda)\}.$$

Statements (iii) and (iv) now follow from Lemma 2.8.

Remark 6.31 The orders of the poles and zeros at infinity of $T(s)$ can be derived from either the p_k's or the s_k's. One therefore has a choice to consider either subspaces of Y or subspaces of U. Observe that there is symmetry with respect to $\boldsymbol{y}$ and $\boldsymbol{u}$ in the definition of W^k. It follows from the previous lemma that we have the following results for an I-O system with transfer function $T(s)$:

(a) $\boldsymbol{u}$ processes $\boldsymbol{y}$ if and only if $T(s)$ is left invertible

(b) $\boldsymbol{y}$ is free if and only if $T(s)$ is right invertible.

Remark 6.32 In the discrete time case ($\mathbf{T} = \mathbb{Z}_+$) the spaces W^k correspond to the spaces $W^k(\mathcal{B})$ as defined in Section 5.2. These can be expressed directly in terms of the behavior $\mathcal{B}$: using the notation of Section 5.2, we have

$$
\begin{aligned}
W^k(\mathcal{B}) &= \chi\{\begin{bmatrix} \boldsymbol{y} \\ \boldsymbol{u} \end{bmatrix} \in \mathcal{B}^0 \mid u_0 = u_1 = \cdots = u_{k-1} = 0\} \quad \text{for } k > 0 \\
W^k(\mathcal{B}) &= \chi\{\begin{bmatrix} \boldsymbol{y} \\ \boldsymbol{u} \end{bmatrix} \in \mathcal{B}^0 \mid y_0 = y_1 = \cdots = y_{k-1} = 0\} \quad \text{for } k < 0 \\
W^k(\mathcal{B}) &= \chi\mathcal{B}^0 \quad \text{for } k = 0.
\end{aligned}
$$

◇

In the following subsections we investigate the input-output structure of systems that are given in first-order form. Our first aim will be to characterize the I-O status of a system in terms of the constant matrices. Secondly, we express rank and pole/zero structure at infinity of the transfer function for I-O systems that are given in first-order form. Lemma 6.30 shows that all issues can be treated in the same framework. We will present results for P representations, DP representations and D representations.

6.3.1 Results for P representations

In this subsection we consider the input-output structure of systems that are given in pencil form (F, G, H_y, H_u)

$$
\begin{aligned}
\sigma G\boldsymbol{\xi} &= F\boldsymbol{\xi} \\
\boldsymbol{y} &= H_y\boldsymbol{\xi} \\
\boldsymbol{u} &= H_u\boldsymbol{\xi}.
\end{aligned}
$$

As before, G and F are mappings from Z to X, whereas H_y and H_u are mappings from Z to Y and Z to U, respectively. As a first step, we express the above subspaces $\pi_Y W^k$ and $\pi_U W^k$ in terms of the constant matrices F, G, H_y and H_u of a P representation (F, G, H_y, H_u). For this, we recall the two iterations (6.3) and (6.38) that were introduced in the previous sections:

$$\mathcal{Z}^0 = Z, \quad \mathcal{Z}^{m+1} = F^{-1}G\mathcal{Z}^m, \tag{6.63}$$

$$\mathcal{N}^0 = \{0\}, \quad \mathcal{N}^{m+1} = G^{-1}F[\mathcal{N}^m \cap \ker H_y \cap \ker H_u]. \tag{6.64}$$

Let us introduce, in addition, the following iterations:

$$\mathcal{V}_y^0 = \mathcal{Z}^*, \quad \mathcal{V}_y^{m+1} = F^{-1}G\mathcal{V}_y^m \cap \ker H_y, \tag{6.65}$$

$$\mathcal{T}_y^0 = \mathcal{N}^*, \quad \mathcal{T}_y^{m+1} = G^{-1}F[\mathcal{T}_y^m \cap \ker H_y], \tag{6.66}$$

$$\mathcal{V}_u^0 = \mathcal{Z}^*, \quad \mathcal{V}_u^{m+1} = F^{-1}G\mathcal{V}_u^m \cap \ker H_u, \tag{6.67}$$

$$\mathcal{T}_u^0 = \mathcal{N}^*, \quad \mathcal{T}_u^{m+1} = G^{-1}F[\mathcal{T}_u^m \cap \ker H_u]. \tag{6.68}$$

Note that there is a difference between the iteration (6.65) and the iteration

$$\bar{\mathcal{V}}_y^0 = Z, \quad \bar{\mathcal{V}}_y^{m+1} = F^{-1}G\bar{\mathcal{V}}_y^m \cap \ker H_y, \tag{6.69}$$

which coincides with the iteration (2.24) for the pencil

$$\begin{bmatrix} sG - F \\ H_y \end{bmatrix}. \tag{6.70}$$

However, the limit spaces of the two iterations coincide by Lemma 2.35. In case that $G\mathcal{Z}^* = \operatorname{im} G$, a stronger statement can be made: then $\mathcal{V}_y^k = \bar{\mathcal{V}}_y^k$ for all $k \geq 1$. In a similar way we might compare the iteration (6.66) and the iteration

$$\bar{\mathcal{T}}_y^0 = \{0\}, \quad \bar{\mathcal{T}}_y^{m+1} = G^{-1}F[\bar{\mathcal{T}}_y^m \cap \ker H_y], \tag{6.71}$$

which coincides with the iteration (2.29) for the pencil (6.70). The limit spaces of these two iterations coincide as well, as can be shown by a similar argument. In case that $\mathcal{N}^* = \ker G$, a stronger statement can be made: then $\mathcal{T}_y^k = \bar{\mathcal{T}}_y^{k+1}$ for all $k \geq 0$. Analogous remarks can be made for the iterations (6.67) and (6.68) (interchange y and u).

We are now ready to express the above subspaces $\pi_Y W^k$ and $\pi_U W^k$ in terms of F, G, H_y and H_u.

Lemma 6.33 Let $\Sigma = (\mathbf{T}, Y \oplus U, \mathcal{B}) \in \mathcal{L}^{p+m}$ and let $\mathcal{B}$ be given by a P representation (F, G, H_y, H_u). Let $\mathcal{Z}^*$ and $\mathcal{N}^*$ be defined as the limit spaces of the iterations (6.63) and (6.64), respectively. Let $\mathcal{V}_y^k$, $\mathcal{T}_y^k$, $\mathcal{V}_u^k$ and $\mathcal{T}_u^k$ be given by the iterations (6.65), (6.66), (6.67) and (6.68), respectively. Then the following hold:

(i) $\pi_Y W^k = H_y[\mathcal{V}_u^k \cap \mathcal{N}^*]$ for all $k \geq 0$

(ii) $\pi_U W^k = H_u[\mathcal{V}_y^{-k} \cap \mathcal{N}^*]$ for all $k \leq 0$

(iii) $\pi_Y W^k = H_y[\mathcal{T}_y^{-k} \cap \mathcal{Z}^*]$ for all $k \leq 0$

(iv) $\pi_U W^k = H_u[\mathcal{T}_u^k \cap \mathcal{Z}^*]$ for all $k \geq 0$.

Proof We first prove (i). Let $y \in \pi_Y W^k$. Then there exist $y(\lambda) \in \lambda^{-1}Y_\infty(\lambda)$ and $u(\lambda) \in \lambda^{-1}U_\infty(\lambda)$ such that $y = \pi_1 y(\lambda)$ and (Lemma 3.16)

$$\begin{bmatrix} y(\lambda) \\ \lambda^{-k}u(\lambda) \end{bmatrix} \in \begin{bmatrix} H_y \\ H_u \end{bmatrix} [\ \ker\ (\lambda G - F)].$$

Let $\xi(\lambda) \in Z(\lambda)$ be such that

$$(\lambda G - F)\xi(\lambda) = 0 \tag{6.72}$$
$$y(\lambda) = H_y\xi(\lambda) \tag{6.73}$$
$$\lambda^{-k}u(\lambda) = H_u\xi(\lambda). \tag{6.74}$$

Write $\xi(\lambda)$ as

$$\xi(\lambda) = \xi_{-l}\lambda^l + \xi_{-l+1}\lambda^{l-1} + \cdots + \xi_0 + \xi_1\lambda^{-1}\cdots.$$

It follows from (6.72) that $G\xi_{-l} = 0$ and $G\xi_{i+1} = F\xi_i$ for all $i \geq -l$, so that $\xi_{-l} \in \mathcal{N}^1$ and $\xi_i \in \mathcal{Z}^*$ for all $i \geq -l$. From (6.73–6.74) it follows that $y = H_y\xi_1$, $H_y\xi_i = 0$ for $-l \leq i \leq 0$ and $H_u\xi_i = 0$ for $-l \leq i \leq k$. As a result, $\xi_1 \in \mathcal{V}_u^k$ and also $\xi_1 \in \mathcal{N}^{l+2}$, so certainly $\xi_1 \in \mathcal{N}^*$. This proves that $\pi_Y W^k \subset H_y[\mathcal{V}_u^k \cap \mathcal{N}^*]$. Conversely, let $y = H_y z$ with $z \in \mathcal{V}_u^k \cap \mathcal{N}^*$. Since $z \in \mathcal{N}^*$, there exist $\xi_0, \xi_{-1}, \ldots, \xi_{-l}$ such that $G\xi_{-l} = 0$ and

$$Gz = F\xi_0,\ G\xi_0 = F\xi_{-1}, \ldots, G\xi_{-l+1} = F\xi_l$$

while

$$H_y\xi_0 = H_y\xi_{-1} = \cdots = H_y\xi_{-l} = 0$$

and also

$$H_u\xi_0 = H_u\xi_{-1} = \cdots = H_u\xi_{-l} = 0.$$

Furthermore, $z \in \mathcal{V}_u^k$ implies that there exist $\xi_2, \xi_3, \ldots, \xi_{k+1}$ such that $\xi_{k+1} \in \mathcal{Z}^*$ and

$$Fz = G\xi_2,\ F\xi_2 = G\xi_3, \ldots, F\xi_k = G\xi_{k+1}$$

while

$$H_u z = H_u\xi_2 = \cdots = H_u\xi_k = 0$$

(take $\xi_{k+1} = z$ if $k = 0$). As a result, we have that $(\lambda G - F)(z(\lambda) + \xi_{k+1}\lambda^{-k-1}) = -(F\xi_{k+1})\lambda^{-k-1}$, where

$$z(\lambda) = \xi_{-l}\lambda^l + \xi_{-l+1}\lambda^{l-1} + \cdots + \xi_0 + z\lambda^{-1} + \xi_2\lambda^{-2} + \cdots + \xi_k\lambda^{-k}.$$

We may now use the fact that $\xi_{k+1} \in \mathcal{Z}^*$ and apply Lemma 2.29 (use Remark 2.30). It follows that there exists a strictly proper rational function $\tilde{\xi}(\lambda)$ such that $(\lambda G - F)\tilde{\xi}(\lambda) = F\xi_{k+1}$. Now define

$$\xi(\lambda) := z(\lambda) + (\xi_{k+1} + \tilde{\xi}(\lambda))\lambda^{-k-1}.$$

Then $y = \pi_1 H_y\xi(\lambda)$, $(\lambda G - F)\xi(\lambda) = 0$ and both $H_y\xi(\lambda)$ and $\lambda^{-k}H_u\xi(\lambda)$ are strictly proper. As a result,

$$\begin{bmatrix} H_y\xi(\lambda) \\ \lambda^{-k}H_u\xi(\lambda) \end{bmatrix} \in \mathcal{W}_*$$

so that $y \in \pi_Y W^k$. We conclude that $\pi_Y W^k \supset H_y[\mathcal{V}_u^k \cap \mathcal{N}^*]$, and this proves (i). By interchanging y and u, we get (ii). Statements (iii) and (iv) are proven in a similar way. ◇

We now come to the three main theorems of this subsection.

Theorem 6.34 Let $\Sigma = (\mathbf{T}, Y \oplus U, \mathcal{B}) \in \mathcal{L}^{p+m}$ and let $\mathcal{B}$ be given by a P representation (F, G, H_y, H_u). Let $\mathcal{Z}^*$ and $\mathcal{N}^*$ be the limit spaces of the iterations (6.63) and (6.64), respectively. Let $\bar{\mathcal{V}}_u^*$ be the limit space of

$$\bar{\mathcal{V}}_u^0 = Z, \quad \bar{\mathcal{V}}_u^{m+1} = F^{-1}G\bar{\mathcal{V}}_u^m \cap \ker H_u \tag{6.75}$$

and let $\bar{\mathcal{T}}_u^*$ be the limit space of

$$\bar{\mathcal{T}}_u^0 = \{0\}, \quad \bar{\mathcal{T}}_u^{m+1} = G^{-1}F[\bar{\mathcal{T}}_u^m \cap \ker H_u]. \tag{6.76}$$

Then the following hold:

(i) $\boldsymbol{y}$ processes $\boldsymbol{u}$ if and only if $H_y[\bar{\mathcal{V}}_u^* \cap \mathcal{N}^*] = \{0\}$

(ii) $\boldsymbol{u}$ is free if and only if $H_u[\bar{\mathcal{T}}_u^* \cap \mathcal{Z}^*] = U$.

Proof By Lemma 2.35 the space $\bar{\mathcal{V}}_u^*$ coincides with the limit space of the iteration (6.67). In an analogous way the space $\bar{\mathcal{T}}_u^*$ coincides with the limit space of the iteration (6.68). The statements in the theorem now follow immediately from Lemma 6.33.

Theorem 6.35 Let $\Sigma = (\mathbf{T}, Y \oplus U, \mathcal{B}) \in \mathcal{L}^{p+m}$ be an I-O system with transfer function $T(s)$. Let $\mathcal{Z}^*$, $\mathcal{N}^*$, $\bar{\mathcal{V}}_y^*$ and $\bar{\mathcal{T}}_y^*$ be the limit spaces of the iterations (6.63), (6.64), (6.69) and (6.71), respectively. Then the following hold:

(i) dim ker $T(s) =$ dim $H_u[\bar{\mathcal{V}}_y^* \cap \mathcal{N}^*]$

(ii) rank $T(s) =$ dim $H_y[\bar{\mathcal{T}}_y^* \cap \mathcal{Z}^*]$.

Proof As in the proof of the previous theorem, the space $\bar{\mathcal{V}}_y^*$ coincides with the limit space of the iteration (6.65), whereas the space $\bar{\mathcal{T}}_y^*$ coincides with the limit space of the iteration (6.66). The statements now follow by combining Lemma 6.30 and Lemma 6.33.

Theorem 6.36 Let $\Sigma = (\mathbf{T}, Y \oplus U, \mathcal{B}) \in \mathcal{L}^{p+m}$ be an I-O system with transfer function $T(s)$. Let $\mathcal{B}$ be given by a P representation (F, G, H_y, H_u). Let $\mathcal{Z}^*$ and $\mathcal{N}^*$ be the limit spaces of the iterations (6.63) and (6.64), respectively. Let $\mathcal{V}_y^k$, $\mathcal{T}_y^k$, $\mathcal{V}_u^k$ and $\mathcal{T}_u^k$ be given by (6.65), (6.66), (6.67) and (6.68), respectively. Let us denote the number of poles at ∞ of $T(s)$ of order $\geq k$ ($\leq k$) by p_k (s_k) ($k \in \mathbb{Z}$). Then the following hold:

(i) $p_k = \dim H_y[\mathcal{V}_u^k \cap \mathcal{N}^*]$ for all $k \geq 0$

(ii) $p_k = \dim H_y[\mathcal{T}_y^{-k} \cap \mathcal{Z}^*]$ for all $k \leq 0$

(iii) $s_k = \dim H_u[\mathcal{T}_u^k \cap \mathcal{Z}^*] - \dim \ker T(s)$ for all $k \geq 0$

(iv) $s_k = \dim H_u[\mathcal{V}_y^{-k} \cap \mathcal{N}^*] - \dim \ker T(s)$ for all $k \leq 0$.

Proof The statements follow by combining Lemma 6.30 and Lemma 6.33.

Remark 6.37 It is easily checked (from Definition 3.10) that $T(s)$ is proper if and only if the DV space W^0 satisfies

$$Y \cap W^0 = \{0\}.$$

On the other hand, it follows from Lemma 6.33 that W^0 is given by

$$W^0 = \begin{bmatrix} H_y \\ H_u \end{bmatrix} [\mathcal{Z}^* \cap \mathcal{N}^*]$$

so that we conclude that $T(s)$ is proper if and only if

$$H_y[\mathcal{Z}^* \cap \mathcal{N}^* \cap \ker H_u] = \{0\}. \tag{6.77}$$

Indeed, (6.77) coincides with $p_1 = 0$ in the above theorem. ◇

Below we want to compare our geometric formulas with expressions in terms of matrix pencils. For this, we need geometric formulas that describe the rank and the pole/zero structure at infinity of an arbitrary matrix pencil $sE - A$. The next theorem shows that such formulas can be obtained from Theorem 6.35 and Theorem 6.36 in an elegant way. It should be noted, however, that geometric formulas for the pole/zero structure at infinity and rank of matrix pencils have been derived before; see e.g. [4, 6, 13, 42].

Let us consider the matrix $sE - A$; as before E and A are mappings from X to Z. We recall the basic iterations (2.24) and (2.29) that were introduced in Section 2.6:

$$V^0 = X, \quad V^{m+1} = A^{-1}EV^m \tag{6.78}$$

and

$$T^0 = \{0\}, \quad T^{m+1} = E^{-1}AT^m. \tag{6.79}$$

Theorem 6.38 Let E, $A : X \mapsto Z$ be linear mappings. Denote the number of zeros at infinity of $sE - A$ of order $\leq k$ ($\geq k$) by t_k (j_k) ($k \in \mathbb{Z}$). Then we have

(i) $\dim \ker (sE - A) = \dim (V^* \cap \ker E)$

(ii) $\text{rank } (sE - A) = \dim (AT^* + \text{ im } E)$

(iii) $j_k = \dim (V^k \cap \ker E) - \dim (V^* \cap \ker E)$ for all $k \geq 0$

(iv) $t_k = \dim (AT^{k+1} + \text{ im } E)$ for all $k \geq -1$

(v) $t_k = 0$ for all $k \leq -2$.

Proof Consider the I-O system $\Sigma = (\mathbf{T}, W = Y \oplus U, \mathcal{B})$ with $Y = Z$ and $U = X$ whose behavior $\mathcal{B}$ is given by the AR representation

$$\boldsymbol{y} - (\sigma E - A)\boldsymbol{u} = 0. \tag{6.80}$$

It is obvious that the transfer function $T(s)$ of Σ is given by $T(s) = sE - A$ and that (6.80) can be rewritten as the P representation (F, G, H_y, H_u) with

$$G = [E \quad 0], F = [A \quad I], H_y = [0 \quad I], H_u = [I \quad 0].$$

Let $\mathcal{Z}^*$ and $\mathcal{N}^*$ be the limit spaces of the iterations (6.63) and (6.64), respectively. Using (6.4) and (6.40) respectively, we have

$$G\mathcal{Z}^* = \text{ im } G = \text{ im } E$$

and

$$\mathcal{N}^* = \ker G = \ker E \oplus Z.$$

Let $\mathcal{V}_y^k$, $\mathcal{T}_y^k$, $\mathcal{V}_u^k$ and $\mathcal{T}_u^k$ be given by the iterations (6.65), (6.66), (6.67) and (6.68), respectively. It is easily verified that

$$\begin{aligned}
\mathcal{V}_y^k &= V^k \oplus \{0\} \text{ for all } k \geq 1, \\
\mathcal{T}_y^k &= T^{k+1} \oplus Z \text{ for all } k \geq 0, \\
\mathcal{V}_u^1 &= \{0\} \oplus \text{ im } E, \\
\mathcal{V}_u^k &= \{0\} \oplus \{0\} \text{ for all } k \geq 2, \\
\mathcal{T}_u^0 &= \ker E \oplus Z, \\
\mathcal{T}_u^k &= X \oplus Z \text{ for all } k \geq 1.
\end{aligned}$$

Application of Theorem 6.35 and Theorem 6.36 yields the desired results. ◇

Because of the above theorem, we may express the formulas in Theorem 6.34, Theorem 6.35 and Theorem 6.36 completely in terms of properties of underlying matrix pencils. As in the two previous sections of this chapter, the only restriction is that we have to impose certain nonredundancy conditions. We present our results as five corollaries:

Corollary 6.39 Let $\Sigma = (\mathbf{T}, Y \oplus U, \mathcal{B}) \in \mathcal{L}^{p+m}$ and let $\mathcal{B}$ be given by a P representation (F, G, H_y, H_u) for which

$$\begin{bmatrix} sG - F \\ H_y \\ H_u \end{bmatrix} \text{ has full column rank and has no zeros at infinity.} \tag{6.81}$$

Then $\boldsymbol{y}$ processes $\boldsymbol{u}$ if and only if the matrix

$$\begin{bmatrix} sG - F \\ H_u \end{bmatrix} \tag{6.82}$$

has full column rank.

Proof By Lemma 2.36 the condition (6.81) is equivalent to the following condition:

$$F^{-1}[\text{ im } G] \cap \ker G \cap \ker H_y \cap \ker H_u = \{0\}. \tag{6.83}$$

By the same lemma, it follows from (6.81) that $\mathcal{N}^* = \ker G$. Let $\bar{\mathcal{V}}_u^*$ be the limit space of the iteration (6.75). Because of (6.83) and the fact that $\bar{\mathcal{V}}_u^* \subset F^{-1}[\text{ im } G] \cap \ker H_u$, we have

$$\dim H_y[\bar{\mathcal{V}}_u^* \cap \ker G] = \dim (\bar{\mathcal{V}}_u^* \cap \ker G).$$

It now follows from Theorem 6.34 that $\boldsymbol{y}$ processes $\boldsymbol{u}$ if and only if

$$\bar{\mathcal{V}}_u^* \cap \ker G = \{0\}. \tag{6.84}$$

Observing that the iteration (6.75) coincides with the iteration (2.24) for the pencil (6.82), we conclude from Theorem 6.38 that (6.84) holds if and only if (6.82) has full column rank.

Corollary 6.40 Let $\Sigma = (\mathbf{T}, Y \oplus U, \mathcal{B}) \in \mathcal{L}^{p+m}$ and let $\mathcal{B}$ be given by a P representation (F, G, H_y, H_u) for which

$$sG - F \text{ has full row rank and has no zeros at infinity.} \tag{6.85}$$

Then $\boldsymbol{u}$ is free if and only if the matrix (6.82) has full row rank.

Proof Let $\mathcal{Z}^*$ be the limit space of the iteration (6.63). By Lemma 2.37 the condition (6.85) is equivalent to the following condition:

$$F[\ker G] + \operatorname{im} G = X. \tag{6.86}$$

By the same lemma, it follows from (6.85) that $G\mathcal{Z}^* = \operatorname{im} G$ and in particular that $\mathcal{Z}^* = F^{-1}[\operatorname{im} G]$. Let $\bar{\mathcal{T}}_u^*$ be the limit space of the iteration (6.76). Because of (6.86) and the fact that $\ker G \subset \bar{\mathcal{T}}_u^*$, it follows that

$$F\bar{\mathcal{T}}_u^* + \operatorname{im} G = X.$$

It now follows as in the proof of Corollary 6.3 that

$$\dim H_u[\bar{\mathcal{T}}_u^* \cap F^{-1}[\operatorname{im} G]] + \dim X = \dim \left(\begin{bmatrix} F \\ H_u \end{bmatrix} \bar{\mathcal{T}}_u^* + \operatorname{im} \begin{bmatrix} G \\ 0 \end{bmatrix} \right). \tag{6.87}$$

By Theorem 6.34 $\boldsymbol{u}$ is free if and only if

$$H_u[\bar{\mathcal{T}}_u^* \cap F^{-1}[\operatorname{im} G]] = U$$

so that $\boldsymbol{u}$ is free if and only if the right-hand side of (6.87) equals $\dim U + \dim X$. Observing that the iteration (6.76) coincides with the iteration (2.29) for the pencil (6.82), we conclude from Theorem 6.38 that $\boldsymbol{u}$ is free if and only if (6.82) has full row rank.

Example 6.41 Consider the controller of Example 6.28, which interconnects two mass-spring-damper systems. A minimal P representation is given by (F, G, H_y, H_u), where $F, G : Z \mapsto X$ with $X = \{0\}$ and

$$H_y = \begin{bmatrix} 1 & 0 \\ 1 & 0 \end{bmatrix}, \quad H_u = \begin{bmatrix} 0 & 1 \\ 0 & -1 \end{bmatrix}.$$

The conditions (6.81) and (6.85) are satisfied because of the minimality of the representation. Using Corollary 6.39, we conclude that $\boldsymbol{y}$ does not process $\boldsymbol{u}$ since

$$\begin{bmatrix} sG - F \\ H_u \end{bmatrix}$$

does not have full column rank. Since the above matrix does not have full row rank either, it follows from Corollary 6.40 that $\boldsymbol{u}$ is not free. Note that this corresponds with observations in Example 6.28.

Corollary 6.42 Let $\Sigma = (\mathbf{T}, Y \oplus U, \mathcal{B}) \in \mathcal{L}^{p+m}$ be an I-O system with transfer function $T(s)$. Let $\mathcal{B}$ be given by a P representation (F, G, H_y, H_u) for which (6.81) holds. Then

$$\dim \ker T(s) = \dim \ker \begin{bmatrix} sG - F \\ H_y \end{bmatrix}.$$

Proof See the proof of Corollary 6.39; interchange y and u and use Theorem 6.35 instead of Theorem 6.34.

Corollary 6.43 Let $\Sigma = (\mathbf{T}, Y \oplus U, \mathcal{B}) \in \mathcal{L}^{p+m}$ be an I-O system with transfer function $T(s)$. Let $\mathcal{B}$ be given by a P representation (F, G, H_y, H_u) for which (6.85) holds. Then

$$\text{rank } T(s) = \text{rank} \begin{bmatrix} sG - F \\ H_y \end{bmatrix} - \dim X.$$

Proof See the proof of Corollary 6.40; interchange y and u and use Theorem 6.35 instead of Theorem 6.34.

Corollary 6.44 Let $\Sigma = (\mathbf{T}, Y \oplus U, \mathcal{B}) \in \mathcal{L}^{p+m}$ with $\mathcal{B}$ given by a P representation (F, G, H_y, H_u) for which (6.81) and (6.85) hold. Then Σ is an I-O system if and only if the matrix

$$\begin{bmatrix} sG - F \\ H_u \end{bmatrix} \tag{6.88}$$

is invertible. Moreover, in this case the following hold for the transfer function $T(s)$:

(i) $\dim \ker T(s) = \dim \ \ker \begin{bmatrix} sG - F \\ H_y \end{bmatrix}$

(ii) the number of poles at infinity of order k ($k \geq 1$) of $T(s)$ equals the number of zeros at infinity of order k of (6.88)

(iii) the number of zeros at infinity of order k ($k \geq 1$) of $T(s)$ equals the number of zeros at infinity of order k of the matrix

$$\begin{bmatrix} sG - F \\ H_y \end{bmatrix}. \tag{6.89}$$

In particular, $T(s)$ is proper if and only if $F^{-1}[\text{ im } G] \cap \ker G \cap \ker H_u = \{0\}$.

Proof The first statement of the corollary was already proven in Corollary 6.39 and Corollary 6.40. Statement (i) follows from Corollary 6.42. As in the proofs of Corollary 6.39 and Corollary 6.40, it follows from (6.81) and (6.85) that (6.83) and (6.86) hold and that $\mathcal{N}^* = \ker G$ and $G\mathcal{Z}^* = \text{im } G$. In particular we have that $\mathcal{Z}^* = F^{-1}[\text{ im } G]$. To prove (ii), let $\mathcal{V}_u^k$ be given by the iteration (6.67). Note that $\mathcal{V}_u^k \subset F^{-1}[\text{ im } G] \cap \ker H_u$ for all $k \geq 1$. It now follows from (6.83) that, for all $k \geq 1$,

$$\dim H_y[\mathcal{V}_u^k \cap \ker G] = \dim (\mathcal{V}_u^k \cap \ker G).$$

Since $G\mathcal{Z}^* = \text{ im } G$, we also have, for all $k \geq 1$, that $\mathcal{V}_u^k = \bar{\mathcal{V}}_u^k$, where $\bar{\mathcal{V}}_u^k$ is given by the iteration (6.75). Consequently, we conclude from Theorem 6.36 that, for all $k \geq 1$,

$$p_k = \dim (\bar{\mathcal{V}}_u^k \cap \ker G),$$

where p_k denotes the number of poles at ∞ of order $\geq k$ of $T(s)$. Denoting the number of poles at ∞ of order k of $T(s)$ by π_k, we then have, for all $k \geq 1$,

$$\pi_k = \dim (\bar{\mathcal{V}}_u^k \cap \ker G) - \dim (\bar{\mathcal{V}}_u^{k+1} \cap \ker G). \tag{6.90}$$

Observing that the iteration (6.75) coincides with the iteration (2.24) for the pencil (6.88), we conclude from Theorem 6.38 that the right-hand side of (6.90) equals the number of zeros at ∞ of order k of the matrix (6.88), and this proves (ii).

To prove (iii), let $\mathcal{T}_y^k$ be given by the iteration (6.66). Note that $\ker G \subset \mathcal{T}_y^k$ for all $k \geq 0$. It now follows as in the proof of Corollary 6.40 that, for all $k \geq 0$,

$$\dim H_y[\mathcal{T}_y^k \cap F^{-1}[\text{ im } G]] = \dim \left(\begin{bmatrix} F \\ H_y \end{bmatrix} \mathcal{T}_y^k + \text{ im } \begin{bmatrix} G \\ 0 \end{bmatrix}\right) - \dim X.$$

Since $\mathcal{N}^* = \ker G$, we also have, for all $k \geq 0$, that $\mathcal{T}_y^k = \bar{\mathcal{T}}_y^{k+1}$, where $\bar{\mathcal{T}}_y^k$ is given by the iteration (6.71). Consequently, we conclude from Theorem 6.36 that, for all $k \leq 0$,

$$p_k = \dim \left(\begin{bmatrix} F \\ H_y \end{bmatrix} \bar{\mathcal{T}}_y^{-k+1} + \text{ im } \begin{bmatrix} G \\ 0 \end{bmatrix}\right) - \dim X$$

so that, for all $k \leq -1$,

$$\pi_k = \dim \left(\begin{bmatrix} F \\ H_y \end{bmatrix} [\bar{\mathcal{T}}_y^{-k+1}] + \text{ im } \begin{bmatrix} G \\ 0 \end{bmatrix}\right)$$
$$-\dim \left(\begin{bmatrix} F \\ H_y \end{bmatrix} [\bar{\mathcal{T}}_y^{-k}] + \text{ im } \begin{bmatrix} G \\ 0 \end{bmatrix}\right). \tag{6.91}$$

Observing that the iteration (6.71) coincides with the iteration (2.29) for the pencil (6.89), we conclude from Theorem 6.38 that the right-hand side of (6.91) equals the number of zeros at ∞ of order k of the matrix (6.89), and this proves (iii). If the matrix (6.88) is invertible, then it has no zeros at infinity if and only if $F^{-1}[\text{ im } G] \cap \ker G \cap \ker H_u = \{0\}$ (Lemma 2.36). The last statement therefore follows from (ii). ◇

We conclude this subsection with the following remark:

Remark 6.45 If the matrix (6.82) is invertible, there does indeed exist a unique transfer function $T(s)$: let $[L_1(s) \quad L_2(s)]$ be the inverse of (6.82) so that

$$\begin{bmatrix} sG - F \\ H_u \end{bmatrix} [L_1(s) \quad L_2(s)] = I.$$

According to Lemma 3.16, the RC space $\mathcal{C}$ is given as

$$\mathcal{C} = \begin{bmatrix} H_y \\ H_u \end{bmatrix} (\ \ker\ (sG - F)), \tag{6.92}$$

and it is easily checked that the right-hand side of (6.92) equals

$$\operatorname{im}\ (\begin{bmatrix} H_y L_2(s) \\ I \end{bmatrix})$$

so that $H_y L_2(s)$ is the transfer function of the system.

6.3.2 Results for DP representations

In this subsection we consider the input-output structure of systems that are given by a DP representation (K, L, M_y, M_u):

$$\sigma K \boldsymbol{\xi} = L \boldsymbol{\xi} + M_y \boldsymbol{y} + M_u \boldsymbol{u}.$$

As before, K and L are mappings from X to Z, M_y is a mapping from Y to Z, and M_u is a mapping from U to Z. There are two ways to obtain results for DP representations as corollaries of results in the previous subsection. The first method is based on the duality that exists under weak external equivalence between DP representations and P representations (Lemma 3.26 and Lemma 6.29). The second method proceeds as in Subsection 6.1.2 by rewriting the DP representation (K, L, M_y, M_u) as the P representation (F, G, H_y, H_u) with

$$G = [K \quad 0 \quad 0], \quad F = [L \quad M_y \quad M_u],$$

$$H_y = [0 \quad I \quad 0], \quad H_u = [0 \quad 0 \quad I]. \tag{6.93}$$

Formulas concerning the input-output structure of a system in dual pencil form can then be derived from Theorem 6.34, Theorem 6.35 and Theorem 6.36 in the same way as a formula concerning the observability indices of a system in dual pencil form (Theorem 6.5) was derived from Theorem 6.1. Here we will not spell out these results explicitly. Instead we formulate results concerning the input-output structure of a system in dual pencil form under the nonredundancy conditions that correspond to (6.81) and (6.85), respectively, through the relation (6.93). The nonredundancy condition that corresponds to (6.81) is

$$sK - L \text{ has full column rank and has no zeros at infinity,} \tag{6.94}$$

whereas the nonredundancy condition that corresponds to (6.85) is

$$\begin{bmatrix} sK - L & M_y & M_u \end{bmatrix} \text{ has full row rank and has no zeros at infinity.} \quad (6.95)$$

The following five theorems are immediate from Corollary 6.39, Corollary 6.40, Corollary 6.42, Corollary 6.43 and Corollary 6.44, respectively.

Theorem 6.46 Let $\Sigma = (\mathbf{T}, Y \oplus U, \mathcal{B}) \in \mathcal{L}^{p+m}$ and let $\mathcal{B}$ be given by a DP representation (K, L, M_y, M_u) for which (6.94) holds. Then $\boldsymbol{y}$ processes $\boldsymbol{u}$ if and only if the matrix $[sK - L \quad M_y]$ has full column rank.

Theorem 6.47 Let $\Sigma = (\mathbf{T}, Y \oplus U, \mathcal{B}) \in \mathcal{L}^{p+m}$ and let $\mathcal{B}$ be given by a DP representation (K, L, M_y, M_u) for which (6.95) holds. Then $\boldsymbol{u}$ is free if and only if the matrix $[sK - L \quad M_y]$ has full row rank.

Example 6.48 Consider the controller of Example 6.28, which interconnects two mass-spring-damper systems. The representation (6.58) is in fact a minimal DP representation (K, L, M_y, M_u), where $K, L : X \mapsto Z$ with $X = \{0\}$ and

$$M_y = \begin{bmatrix} -1 & 1 \\ 0 & 0 \end{bmatrix}, \quad M_u = \begin{bmatrix} 0 & 0 \\ -1 & -1 \end{bmatrix}.$$

The conditions (6.94) and (6.95) are satisfied because of the minimality of the representation. Using Theorem 6.46, we conclude that $\boldsymbol{y}$ does not process $\boldsymbol{u}$ since $[sK - L \quad M_y]$ does not have full column rank. Since this matrix does not have full row rank either, it follows from Theorem 6.47 that $\boldsymbol{u}$ is not free. Note that this corresponds with observations in Example 6.28.

Theorem 6.49 Let $\Sigma = (\mathbf{T}, Y \oplus U, \mathcal{B}) \in \mathcal{L}^{p+m}$ be an I-O system with transfer function $T(s)$, and let its behavior $\mathcal{B}$ be given by a DP representation (K, L, M_y, M_u) for which (6.94) holds. Then

$$\dim \ \ker T(s) = \dim \ \ker \ [sK - L \quad M_u].$$

Theorem 6.50 Let $\Sigma = (\mathbf{T}, Y \oplus U, \mathcal{B}) \in \mathcal{L}^{p+m}$ be an I-O system with transfer function $T(s)$. Let its behavior $\mathcal{B}$ be given by a DP representation (K, L, M_y, M_u) for which (6.95) holds. Then

$$\operatorname{rank} T(s) = \operatorname{rank} \ [sK - L \quad M_u] + \dim Y - \dim Z.$$

Theorem 6.51 Let $\Sigma = (\mathbf{T}, Y \oplus U, \mathcal{B}) \in \mathcal{L}^{p+m}$ with $\mathcal{B}$ given by a DP representation (K, L, M_y, M_u) for which (6.94) and (6.95) hold. Then Σ is an I-O system if and only if the matrix $[sK - L \quad M_y]$ is invertible. Moreover, in this case the following hold for the transfer function $T(s)$:

(i) $\dim \ker T(s) = \dim \ \ker \ [sK - L \quad M_u]$

(ii) the number of poles at infinity of order k ($k \geq 1$) of $T(s)$ equals the number of zeros at infinity of order k of $[sK - L \quad M_y]$

(iii) the number of zeros at infinity of order k ($k \geq 1$) of $T(s)$ equals the number of zeros at infinity of order k of the matrix $[sK - L \quad M_u]$.

In particular, $T(s)$ is proper if and only if $L[\ \ker K] + \ \text{im}\ L + \ \text{im}\ M_y = Z$.

Proof The first statement of the theorem is immediate from Theorem 6.46 and Theorem 6.47, whereas statement (i) follows from Theorem 6.49. To prove (ii) and (iii), we write (K, L, M_y, M_u) as the P representation (F, G, H_y, H_u) with F, G, H_y and H_u defined as in (6.93). Because of (6.94) and (6.95), the conditions in Corollary 6.44 are satisfied, and it follows from this corollary that, for all $k \geq 1$, the number of poles at infinity of order k of $T(s)$ equals the number of zeros at infinity of order k of the matrix

$$\begin{bmatrix} sK - L & -M_y & -M_u \\ 0 & 0 & I \end{bmatrix}. \tag{6.96}$$

Clearly, the zero structure at infinity of (6.96) equals the zero structure at infinity of the matrix

$$\begin{bmatrix} sK - L & -M_y & 0 \\ 0 & 0 & I \end{bmatrix} \tag{6.97}$$

that is obtained from left multiplication of (6.96) by the $\mathbb{R}_\infty(s)$-unimodular matrix

$$\begin{bmatrix} I & M_u \\ 0 & I \end{bmatrix}.$$

If $k \geq 1$, the number of zeros at infinity of order k of (6.97) evidently equals the number of zeros at infinity of order k of $[sK - L \quad M_y]$, and this proves (ii). Statement (iii) follows in a similar way. If the matrix $[sK - L \quad M_y]$ is invertible, then it has no zeros at infinity if and only if $L[\ \ker K] + \ \text{im}\ L + \ \text{im}\ M_y = Z$ (Lemma 2.37). The last statement therefore follows from (ii).

6.3.3 Results for D representations

In this section we investigate the input-output structure of a system that is given by a D representation (E, A, B, C, D):

$$\begin{aligned} \sigma E \boldsymbol{\xi} &= A\boldsymbol{\xi} + B\boldsymbol{u} \\ \boldsymbol{y} &= C\boldsymbol{\xi} + D\boldsymbol{u}. \end{aligned}$$

As before, E and A are mappings from X_d to X_e, B is a mapping from U to X_e, C is a mapping from Y to X_e, and D is a mapping from U to

Y. Again (see Section 6.1 and Section 6.2) there are two ways to obtain results for D representations as corollaries of previous results. The first method proceeds by rewriting the D representation (E, A, B, C, D) as the P representation (F, G, H_y, H_u) with

$$G = [E \quad 0], \quad F = [A \quad B], \quad H_y = [C \quad D], \quad H_u = [0 \quad I]. \tag{6.98}$$

The second method proceeds by rewriting (E, A, B, C, D) as the DP representation (K, L, M_y, M_u) with

$$K = \begin{bmatrix} E \\ 0 \end{bmatrix}, \quad L = \begin{bmatrix} A \\ -C \end{bmatrix}, \quad M_y = \begin{bmatrix} 0 \\ I \end{bmatrix}, \quad M_u = \begin{bmatrix} B \\ -D \end{bmatrix}. \tag{6.99}$$

Formulas concerning the input-output structure of a system in descriptor form can be derived through the connection (6.98) as corollaries of Theorem 6.34, Theorem 6.35 and Theorem 6.36. This can be done in the same way as a formula concerning the controllability indices of a system in descriptor form (Theorem 6.20) was derived from Theorem 6.14. Here we will not spell out these results explicitly. Instead we formulate results concerning the input-output structure of a system in descriptor form under the nonredundancy conditions that correspond to (6.81) and (6.85), respectively, through the relation (6.98) (or, equivalently, to (6.94) and (6.95), respectively, through the relation (6.99)). It is easily seen that the nonredundancy condition that corresponds to (6.94) is

$$\begin{bmatrix} sE - A \\ C \end{bmatrix} \text{ has full column rank and has no zeros at infinity,,} \tag{6.100}$$

whereas the nonredundancy condition that corresponds to (6.85) is

$$[sE - A \quad B] \text{ has full row rank and has no zeros at infinity.} \tag{6.101}$$

The following five theorems are immediate from Corollary 6.39 (Theorem 6.46), Corollary 6.40 (Theorem 6.47), Corollary 6.42 (Theorem 6.49), Corollary 6.43 (Theorem 6.50), and Corollary 6.44 (Theorem 6.51).

Theorem 6.52 Let $\Sigma = (\mathbf{T}, Y \oplus U, \mathcal{B}) \in \mathcal{L}^{p+m}$ and let $\mathcal{B}$ be given by a D representation (E, A, B, C, D) for which (6.100) holds. Then $\boldsymbol{y}$ processes $\boldsymbol{u}$ if and only if the matrix $sE - A$ has full column rank.

Theorem 6.53 Let $\Sigma = (\mathbf{T}, Y \oplus U, \mathcal{B}) \in \mathcal{L}^{p+m}$ and let $\mathcal{B}$ be given by a D representation (E, A, B, C, D) for which (6.101) holds. Then $\boldsymbol{u}$ is free if and only if the matrix $sE - A$ has full row rank.

Theorem 6.54 Let $\Sigma = (\mathbf{T}, Y \oplus U, \mathcal{B}) \in \mathcal{L}^{p+m}$ be an I-O system with transfer function $T(s)$. Let $\mathcal{B}$ be given by a D representation (E, A, B, C, D) for which (6.100) holds. Then

$$\dim \ \ker T(s) = \dim \ \ker \begin{bmatrix} sE - A & -B \\ C & D \end{bmatrix}.$$

Theorem 6.55 Let $\Sigma = (\mathbf{T}, Y \oplus U, \mathcal{B}) \in \mathcal{L}^{p+m}$ be an I-O system with transfer function $T(s)$. Let $\mathcal{B}$ be given by a D representation (E, A, B, C, D) for which (6.101) holds. Then

$$\text{rank } T(s) = \text{rank} \begin{bmatrix} sE - A & -B \\ C & D \end{bmatrix} - \dim X_e.$$

Theorem 6.56 Let $\Sigma = (\mathbf{T}, Y \oplus U, \mathcal{B}) \in \mathcal{L}^{p+m}$ with $\mathcal{B}$ given by a D representation (E, A, B, C, D) for which (6.100) and (6.101) hold. Then Σ is an I-O system if and only if the matrix $sE - A$ is invertible. Moreover, in this case the following hold for the transfer function $T(s)$:

(i) $\dim \ker T(s) = \dim \ \ker \begin{bmatrix} sE - A & -B \\ C & D \end{bmatrix}$

(ii) the number of poles at infinity of order k $(k \geq 1)$ of $T(s)$ equals the number of zeros at infinity of order k of $sE - A$

(iii) the number of zeros at infinity of order k $(k \geq 1)$ of $T(s)$ equals the number of zeros at infinity of order k of the matrix

$$\begin{bmatrix} sE - A & -B \\ C & D \end{bmatrix}.$$

In particular, $T(s)$ is proper if and only if $A^{-1}[\text{ im } E] \cap \ker E = \{0\}$.

Proof The first statement of the theorem is immediate from Theorem 6.52 and Theorem 6.53, whereas statement (i) follows from Theorem 6.54. The statements (ii) and (iii) follow from Corollary 6.44 (or, equivalently, from Theorem 6.51) through the connection (6.98) ((6.99)). The last statement of the theorem follows from (ii) and Lemma 2.36. ◇

Claims (ii) and (iii) in the above theorem are due to [56, 57, 59], where a completely different proof was given. Thus reformulated, (ii) and (iii) are analogues of well-known results (see [51]) on the pole-zero structure at *finite* points of the complex plane; see the remarks in [59].

Example 6.57 Consider the electrical circuit of Example 1.5, which is given by the D representation (1.11). The representation satisfies conditions (6.100) and (6.101) (use Lemma 2.36 and Lemma 2.37). Consequently, we may apply Theorem 6.56. We find that we are dealing with an I-O system with a proper transfer function $T(s)$ which has one zero at ∞ (use Theorem 6.38 as well) and for which $\dim \ker T(s) = 0$ and $\text{rank } T(s) = 1$. Indeed, we conclude from the equivalent representation (1.12) that

$$T(s) = \frac{r_2}{c r_1 r_2 s + r_1 + r_2}.$$

In case that $r_1 = 0$, a D representation for the system is

$$\begin{bmatrix} cr_2 & 0 \\ 0 & 0 \end{bmatrix}\begin{bmatrix} \dot{\xi}_1 \\ \dot{\xi}_2 \end{bmatrix} = \begin{bmatrix} -1 & r_2 \\ 1 & 0 \end{bmatrix}\begin{bmatrix} \xi_1 \\ \xi_2 \end{bmatrix} + \begin{bmatrix} 0 \\ -1 \end{bmatrix}\boldsymbol{u}$$

$$\boldsymbol{y} = [1 \quad 0]\begin{bmatrix} \xi_1 \\ \xi_2 \end{bmatrix}$$

so that the system has transfer function $T(s) = 1$. On the other hand, $A^{-1}[\text{ im } E] \cap \ker E \neq \{0\}$ so that according to Theorem 6.56, the transfer function cannot be proper. However, Theorem 6.56 cannot be applied since condition (6.100) is not fulfilled.

Example 6.58 Consider the system consisting of the two interconnected electrical circuits of Example 1.6. The system is represented in minimal descriptor form by the representation (1.15). Because of the minimality of the representation, the conditions (6.100) and (6.101) are satisfied. It then follows from Theorem 6.52 that the voltages process the currents, whereas it follows from Theorem 6.53 that the currents are not free. Indeed, the behavior of the voltages is determined by the behavior of the currents, but the behavior of the currents is not arbitrary: the currents are related through $\boldsymbol{I}_1 = c_1\boldsymbol{I}_2$. We conclude that this system is not an I-O system: it does not have a unique transfer function. ◇

For a standard state space representation, i.e., a D representation with $E = I$, the nonredundancy conditions of Theorem 6.56 are automatically fulfilled. It therefore follows without any further assumptions that the transfer function $T(s) = C(sI - A)^{-1}B + D$ satisfies:

(i) $\dim \ker T(s) = \dim \ \ker \begin{bmatrix} sI - A & -B \\ C & D \end{bmatrix}$

(ii) for all $k \geq 1$ the number of zeros at infinity of order k of $T(s)$ equals the number of zeros at infinity of order k of

$$\begin{bmatrix} sI - A & -B \\ C & D \end{bmatrix}. \tag{6.102}$$

Result (i) is in line with the geometric characterization of left invertibility of $T(s)$ as given in [44, Theorem 5]. The connection as given by (ii) with the zeros at infinity of (6.102) was made explicit by Verghese and Kailath [58]. Various geometric formulas for computing the orders of these zeros have been derived in the literature; see e.g. [11] and [48]. The formulas in [48] coincide with the formulas that are obtained by using Theorem 6.38.

Before presenting a procedure by which a a general D representation can be rewritten in standard state space form, we ask ourselves the following question: what conditions should the matrices E, A, B, C and D satisfy so that (E, A, B, C, D) can be rewritten in standard state space form?

Theorem 6.59 Let $\Sigma = (\mathbf{T}, Y \oplus U, \mathcal{B}) \in \mathcal{L}^{p+m}$ be given by a D representation (E, A, B, C, D). Let X^*, N^*, V^* and T^* be the limit spaces of the iterations (6.25), (6.50), (2.24) and (2.29), respectively. Then Σ has a unique transfer function $T(s)$ if and only if the following conditions are satisfied:

(i) $V^* \cap N^* \subset \ker C$

(ii) $\operatorname{im} B \subset AT^* + EV^*$.

Furthermore, $T(s)$ is proper if and only if

$$(A^{-1}EX^*) \cap N^* \subset \ker C.$$

Proof Write (E, A, B, C, D) as the P representation (F, G, H_y, H_u), where F, G, H_y and H_u are defined as in (6.98). Let $\bar{\mathcal{V}}_u^*$ and $\bar{\mathcal{T}}_u^*$ be the limit spaces of the iterations (6.75) and (6.76), respectively. Then $\bar{\mathcal{V}}_u^* = V^* \oplus \{0\}$ and $\bar{\mathcal{T}}_u^* = T^* \oplus U$. The first statement of the theorem then follows from Theorem 6.34. To prove the second statement of the theorem, let $\mathcal{Z}^*$ and $\mathcal{N}^*$ be the limit spaces of the iterations (6.63) and (6.64), respectively. Then $\mathcal{Z}^* = \{(x, u) \in X_d \oplus U \mid Ax + Bu \in EX^*\}$ and $\mathcal{N}^* = N^* \oplus U$. The statement is then immediate from (6.77). ◇

We now give a general procedure for rewriting an I-O system that is given by a D representation in standard state space form. A requirement, of course, is that the conditions of the above theorem are fulfilled. We may choose to avoid checking these conditions: it is possible to follow the procedure below and check the conditions "on the way".

Procedure 6.60 Let a D representation be given by (E, A, B, C, D).

Step 1

Check if conditions (6.100) and (6.101) hold. If they both hold, then proceed with step 2. If at least one of these conditions does not hold, then construct an equivalent D representation that satisfies (6.100) and (6.101) by using the proof of Lemma 4.8: note that conditions (i) and (ii) of Lemma 4.8 are stronger than conditions (6.101) and (6.100), respectively.

Step 2

Check if the matrix $sE - A$ is invertible. If it is invertible, then proceed with step 3. If $sE - A$ is not invertible, then conclude that the system has no transfer function, so the system cannot be given in standard state space form.

Step 3

Check if $A^{-1}[\text{ im } E] \cap \ker E = \{0\}$. If this holds, then proceed with step 4. If it does not hold, then the transfer function of the system is not proper, so the system cannot be given in standard state space form.

Step 4

Decompose the descriptor space X_d as $X_{d1} \oplus X_{d2}$, where $X_{d2} = \ker E$. Decompose the equation space X_e as $X_{e1} \oplus X_{e2}$, where $X_{e1} = \text{im } E$. Accordingly write

$$E = \begin{bmatrix} E_{11} & 0 \\ 0 & 0 \end{bmatrix}, A = \begin{bmatrix} A_{11} & A_{12} \\ A_{21} & A_{22} \end{bmatrix}, B = \begin{bmatrix} B_1 \\ B_2 \end{bmatrix}, C = \begin{bmatrix} C_1 & C_2 \end{bmatrix}.$$

Both E_{11} and A_{22} should now be invertible. The desired standard state space representation is given by

$$\begin{aligned} \sigma \boldsymbol{x} &= E_{11}{}^{-1}(A_{11} - A_{12}A_{22}{}^{-1}A_{21})\boldsymbol{x} + E_{11}{}^{-1}(B_1 - A_{12}A_{22}{}^{-1}B_2)\boldsymbol{u} \\ \boldsymbol{y} &= (C_1 - C_2A_{22}{}^{-1}A_{21})\boldsymbol{x} + (D - C_2A_{22}{}^{-1}B_2)\boldsymbol{u} \end{aligned}$$

Example 6.61 Consider the two interconnected electrical circuits of Example 1.6. To illustrate the above procedure, define $\boldsymbol{u} = \boldsymbol{I}_1 + \boldsymbol{I}_2$ and $\boldsymbol{y} = \boldsymbol{V}_1$. A (nonminimal) D representation is then given by

$$\begin{aligned} \begin{bmatrix} 1 & 0 & 0 \\ 0 & 0 & 0 \\ 0 & 0 & 0 \end{bmatrix} \begin{bmatrix} \dot{\boldsymbol{x}}_1 \\ \dot{\boldsymbol{x}}_2 \\ \dot{\boldsymbol{x}}_3 \end{bmatrix} &= \begin{bmatrix} 0 & 0 & 1 \\ 0 & 1 & -c_1 \\ 0 & 1 & 1 \end{bmatrix} \begin{bmatrix} \boldsymbol{x}_1 \\ \boldsymbol{x}_2 \\ \boldsymbol{x}_3 \end{bmatrix} + \begin{bmatrix} 0 \\ 0 \\ -1 \end{bmatrix} \boldsymbol{u} \\ \boldsymbol{y} &= \begin{bmatrix} 1 & 0 & 0 \end{bmatrix} \begin{bmatrix} \boldsymbol{x}_1 \\ \boldsymbol{x}_2 \\ \boldsymbol{x}_3 \end{bmatrix}. \end{aligned}$$

It is easily checked that (6.100) and (6.101) hold (use Lemma 2.36 and Lemma 2.37). In the terminology of Procedure 6.60 we have

$$E_{11} = 1, \quad A_{22} = \begin{bmatrix} 1 & -c_1 \\ 1 & 1 \end{bmatrix}.$$

Application of the procedure leads to the standard state space representation

$$\begin{aligned} \dot{\boldsymbol{x}} &= \frac{1}{1+c_1}\boldsymbol{u} \\ \boldsymbol{y} &= \boldsymbol{x}. \end{aligned}$$

Example 6.62 Consider the electrical circuit of Example 1.5, which is described by the (nonminimal) D representation (1.11):

$$\begin{aligned} \begin{bmatrix} cr_2 & 0 \\ 0 & 0 \end{bmatrix} \begin{bmatrix} \dot{\xi}_1 \\ \dot{\xi}_2 \end{bmatrix} &= \begin{bmatrix} -1 & r_2 \\ 1 & r_1 \end{bmatrix} \begin{bmatrix} \xi_1 \\ \xi_2 \end{bmatrix} + \begin{bmatrix} 0 \\ -1 \end{bmatrix} \boldsymbol{u} \\ \boldsymbol{y} &= [1 \quad 0] \begin{bmatrix} \xi_1 \\ \xi_2 \end{bmatrix}. \end{aligned}$$

As noted in Example 6.57, conditions (6.100) and (6.101) are satisfied for $r_1 > 0$ and a proper transfer function exists. In the terminology of Procedure 6.60, we have $E_{11} = cr_2$ and $A_{22} = r_1$. Application of the procedure leads to the standard state space representation

$$\begin{aligned} \dot{\boldsymbol{x}} &= -\frac{r_1 + r_2}{cr_1 r_2}\boldsymbol{x} + \frac{1}{cr_1}\boldsymbol{u} \\ \boldsymbol{y} &= \boldsymbol{x}. \end{aligned}$$

◇

7

Conclusions

In this book our main object of study was concerned with several types of first-order representations. We considered so-called P, DP, DZ and D (descriptor) representations in a behavioral setting, regarding them as entities that represent the behavior of linear time-invariant dynamical systems. The fact that the concept of a transfer function has no specific meaning in the behavioral approach enabled us to discuss the most general class of representations: there was no need to put restrictions on the representations. In this way we were able to include representations that pose difficulties to classical transfer function approaches because of their lack of a transfer function. Such representations could, for example, represent controllers that establish an interconnection between system components. We gave an example of such a controller in the context of mass-spring-damper systems.

Questions that arise naturally in this context are the following:

(1) what is the relation between representations that represent the same behavior?

(2) under what conditions is a representation minimal?

(3) how are structural properties of dynamical systems reflected in their representations?

With respect to the first question, we presented realization methods in Chapter 5 for obtaining (minimal) first-order representations of the above types from so-called AR representations. The P representation has, we believe, proved some of its virtue in this context. For minimal first-order representations, the first question was fully answered in Section 4.5, where we derive the constant transformations, according to which equivalent representations (of the same type) are related. In Section 3.4 we showed that the relation between the P representation and the DP representation exhibits duality features.

The second question was dealt with in the first part of Chapter 4 and is, in our opinion, answered in a rather complete way. We have demonstrated that the basic material of Chapter 2, in which results from the algebraic theory are combined with results from the geometric theory, leads to elegant proofs of minimality results.

The third question has been studied in Section 6.1 and Section 6.2 with respect to observability and controllability indices. In Section 6.3 the question is studied with respect to the input-output status of a system; in the same section we also investigate certain integer invariants that are specifically related to input-output systems (rank and poles/zeros at infinity of the transfer function). In these sections geometric formulas are derived in terms of representations with an arbitrary amount of redundancy. In the derivation of the formulas, we found that the P representation lends itself most easily to computation. Intuitively, we ascribe this to the fact that relevant rational spaces are expressed as images of kernels for P representations, whereas they are given as inverse images of images in terms of DP representations. In Section 6.3 we gave a procedure for rewriting a D representation in standard state space form.

We have extended the scope of definition of observability and controllability indices to systems that have a nonproper transfer function and to systems that do not have a transfer function (such as controllers that establish an interconnection). In our definition the controllability indices are invariant under a type of equivalence that is less natural in the present context of C^∞-behaviors/$\mathbb{R}^{\mathbb{Z}_+}$-behaviors (but not so in the context of L_2-behaviors/ℓ_2-behaviors). We have made this type of equivalence explicit in Section 3.3, where it is called "weak external equivalence". We believe to have demonstrated in Section 6.2 that weak external equivalence can be a useful tool for investigating those structural properties of a dynamical system that are invariant under this type of equivalence.

We expect the duality between the P and the DP representations to be important for control theoretic problems. Our expectation is based on the fact that the combination of dual concepts such as left polynomial factorizations and right polynomial factorizations (of transfer functions) has proved powerful in solving control theoretic problems. As a topic of further research we propose to investigate whether the P form and the DP form can be combined in a similar way. Such research could be at the basis of new geometric results in a control theoretic context.

Bibliography

[1] Aplevich, J.D. (1981) : Time-domain input-output representations of linear systems, *Automatica*, Vol. 17, pp. 509-522

[2] Aplevich, J.D. (1985) : Minimal representations of implicit linear systems, *Automatica*, Vol. 21, pp. 259-269

[3] Aplevich, J.D. (1991) : *Implicit Linear Systems*, Springer-Verlag

[4] Armentano, V.A. (1986) : The pencil $sE - A$ and controllability-observability for generalized linear systems: a geometric approach, *SIAM J. Control Optimiz.*, Vol. 24, pp. 616-638

[5] Atiyah, M.F. and I.G. MacDonald (1969) : *Introduction to Commutative Algebra*, Addison-Wesley

[6] Bernhard, P. (1982) : On singular implicit linear dynamical systems, *SIAM J. Control Optimiz.*, Vol. 20, pp. 612-633

[7] Blomberg, H. and R. Ylinen (1983) : *Algebraic Theory for Multivariable Linear Systems*, Ac. Press, London

[8] Campbell, S.L. (1980) : *Singular Systems of Differential Equations*, Part I, Pitman Publ. Inc., San Francisco

[9] Clancey, K. and I.C. Gohberg (1981) : *Factorization of Matrix Functions and Singular Integral Operators*, Birkhäuser, Basel

[10] Clements, D. J. (1992) : *Minimal descriptor forms*, Internal Report, University of New South Wales, Australia

[11] Commault, C. and J.M. Dion (1982) : Structure at infinity of linear multivariable systems: a geometric approach, *IEEE Trans. Aut. Control*, Vol. AC-27, pp. 693-696

[12] Conte, G. and A.M. Perdon (1982) : Generalized state space realizations for nonproper rational transfer functions, *Syst. Control Lett.*, Vol. 1, pp. 270-276

[13] Dieudonné, J. (1946) : Sur la réduction canonique des couples des matrices, *Bull. Soc. Math. France*, Vol. 74, pp. 130-146

[14] Forney, G.D. (1970) : Convolution codes I: Algebraic structure, *IEEE Trans. Information Theory*, Vol. IT-16, pp. 720-738

[15] Forney, G.D. (1975) : Minimal bases of rational vector spaces, with applications to multivariable linear systems, *SIAM J. Control Optimiz.*, Vol. 13, pp. 493-520

[16] Fuhrmann, P.A. (1976) : Algebraic system theory: an analyst's point of view, *J. Franklin Inst.*, Vol. 301, pp. 521-540

[17] Fuhrmann, P.A. and J.C. Willems (1979) : Factorization indices at infinity for rational matrix functions, *Integral Equations and Operator Theory*, Vol. 2/3, pp. 287-301

[18] Fuhrmann, P.A. (1981) : *Linear Systems and Operators in Hilbert Space*, McGraw-Hill, New York

[19] Gantmacher, F.R. (1959) : *Matrix Theory*, Chelsea, New York (Russian original: 1954)

[20] Glüsing-Lüerßen, H. (1990) : Feedback canonical form for singular systems, *Int. J. Control*, Vol. 52, No. 2, pp. 347-376

[21] Grimm, J. (1988) : Realization and canonicity for implicit systems, *SIAM J. Control Optimiz.*, Vol. 26, pp. 1331-1347

[22] Hautus, M.L.J. (1975) : The formal Laplace transform for smooth linear systems, in: *Mathematical Systems Theory* (ed.: G. Marchesini and S.K. Mitter), pp. 29-47, Lect. Notes Econ. Math. Syst., Vol. 131, Springer Verlag, New York (Proc. Int. Symp., Udine, Italy, 1975)

[23] Hautus, M.L.J. (1980) : (A, B)-invariant subspaces and stabilizability subspaces, a frequency domain description, *Automatica*, Vol. 16, pp. 703-707

[24] Hautus, M.L.J. and L.M. Silverman (1983) : System structure and singular control, *Lin. Alg. Appl.*, Vol. 50, pp. 369-402

[25] Janssen, P.H.M. (1988) : General results on the McMillan degree and the Kronecker indices of ARMA and MFD models, *Int. J. Control*, Vol. 48, pp. 591-608

[26] Kailath, T. (1980) : *Linear Systems*, Prentice-Hall,

[27] Kalman, R.E., P.L. Falb, and M.A. Arbib (1969) : *Topics in Mathematical System Theory*, McGraw Hill, New York

[28] Kučera, V. and P. Zagalak (1988) : Fundamental theorem of state feedback for singular systems, *Automatica*, Vol. 24, No. 5, pp. 653-658

[29] Kuijper, M. and J.M. Schumacher (1990) : Realization of autoregressive equations in pencil and descriptor form, *SIAM J. Control Optimiz.*, Vol. 28, pp. 1162-1189

[30] Kuijper, M. and J.M. Schumacher (1991) : Minimality of descriptor representations under external equivalence, *Automatica*, Vol. 27, No. 6, pp. 985-995

[31] Kuijper, M. and J.M. Schumacher (1991) : Realization of finite and infinite modes, *Proc. ISIRS'91 (Second International Symposium on Implicit and Robust Systems)*, pp. 113-116, Warschau, July 1991

[32] Kuijper, M. and J.M. Schumacher (1992) : Realization and partial fractions, *Lin. Alg. Appl.*, Vol. 169, pp. 195-223

[33] Kuijper, M. (1992) : Descriptor representations without direct feedthrough term, *Automatica*, Vol. 28, No. 3, pp. 633-639

[34] Kuijper, M. and J.M. Schumacher (1993) : Input-output structure of linear differential/algebraic systems, *IEEE Trans. Aut. Control*, Vol. AC-38, No. 3, pp. 404-414

[35] Kuijper, M. (1993) : Lag indices of a dynamical system, submitted to *Conference Proceedings MTNS 93*, August 1993

[36] Kuijper, M. (1993) : Why do stabilizing controllers stabilize?, Report W-9402, University of Groningen (submitted to *Automatica* in November 1993)

[37] Kung, S.-Y., B.C. Lévy, M. Morf, and T. Kailath (1977) : New results in 2-D systems theory, Part I, *Proc. of the IEEE*, Vol. 65, No. 6, pp. 945-961

[38] Leontieff, W.W. (1953) : Static and dynamic theory, in: *Studies in the Structure of the American Economy* (ed: W.W. Leontieff), Oxford University Press, New York

[39] Lewis, F.L. (1986) : A survey of linear singular systems, *Circuits Systems Signal Process.*, Vol. 5. pp. 3-36

[40] Luenberger, D.G. and A. Arbel (1977) : Singular dynamic Leontieff systems, *Econometrica*, Vol. 45, pp. 991-995

[41] Luenberger, D.G. (1977) : Dynamic equations in descriptor form, *IEEE Trans. Aut. Control*, Vol. AC-22, pp. 312-321

[42] Malabre, M. (1989) : Generalized linear systems: geometric and structural approaches, *Lin. Alg. Appl.*, Vol. 122/123/124, pp. 591-621

[43] McMillan, B. (1952) : Introduction to formal realization theory, *Bell Syst. Tech. J.*, Vol. 31, pp. 217-279 (part I), pp. 541-600 (part II)

[44] Morse, A.S. and W.M. Wonham (1971) : Status of noninteracting control, *IEEE Trans. Aut. Control*, Vol. AC-16, pp. 568-581

[45] Morse, A.S. (1975) : System invariants under feedback and cascade control, in: *Mathematical Systems Theory* (ed.: G. Marchesini and S.K. Mitter), pp. 61-74, Lect. Notes Econ. Math. Syst., Vol. 131, Springer Verlag, New York (Proc. Int. Symp., Udine, Italy, 1975)

[46] Murota, K. and J.W. van der Woude (1991) : Structure at infinity of structured descriptor systems and its applications, *SIAM J. Cont. Optimiz.*, Vol. 29, no. 4, pp. 878-894

[47] Newman, M. (1972) : *Integral Matrices*, Academic Press, New York

[48] Nijmeijer, H. and J.M. Schumacher (1985), On the inherent integration structure of nonlinear systems, *IMA J. Math. Control Inf.*, Vol. 2, pp. 87-107

[49] Raa, M.H. ten (1986) : Dynamic input-output analysis with distributed activities, *Rev. Econ. Stat.*, Vol. 68, pp. 300-310

[50] Rocha, M.P.M. (1990) : *Structure and Representation of 2-D Systems*, Ph.D.Thesis, University Groningen, The Netherlands

[51] Rosenbrock, H.H. (1970) : *State-space and Multivariable Theory*, Wiley, New York

[52] Rosenbrock, H.H. (1974) : Structural properties of linear dynamical systems, *Int. J. Control*, Vol. 20, pp. 191-202

[53] Schumacher, J.M. (1988) : Transformations of linear systems under external equivalence, *Lin. Alg. Appl.*, Vol. 102, pp. 1-34

[54] Schumacher, J.M. (1991) : A pointwise criterion for controller robustness, *Syst. Control Lett.*, Vol. 18, No. 1, pp. 1-8

[55] Soethoudt, J.M. (1993) : *Introduction to a behavioral approach for continuous-time systems*, Ph.D.Thesis, Technische Universiteit Eindhoven, The Netherlands

[56] Verghese, G.C. (1978) : *Infinite Frequency Behavior in Generalized Dynamical Systems*, Ph.D.Thesis, Information Systems Lab., Stanford University

[57] Verghese, G.C. and T. Kailath (1979) : Impulsive behavior in dynamical systems: structure and significance, *Proceedings 4th MTNS*, Delft, The Netherlands, pp. 162-168

[58] Verghese, G.C. and T. Kailath (1979) : Comments on "On structural invariants and the root-loci of linear multivariable systems", *Int. J. Control*, Vol. 29, pp. 1077-1080

[59] Verghese, G.C., P. van Dooren, and T. Kailath (1979) : Properties of the system matrix of a generalized state-space system, *Int. J. Control*, Vol. 30, pp. 235-243

[60] Verghese, G.C., B.C. Lévy, and T. Kailath (1981) : A generalized state space for singular systems, *IEEE Trans. Aut. Control*, Vol. AC-26, pp. 811-831

[61] Verghese, G.C. and T. Kailath (1981) : Rational matrix structure, *IEEE Trans. Aut. Control*, Vol. AC-26, pp. 434-439

[62] Vidyasagar, M. (1985) : *Control System Synthesis: A Factorization Approach*, MIT Press, Cambridge

[63] Weiland, S. (1991) : *Theory of Approximation and Disturbance Attenuation for Linear Systems*, Ph.D.Thesis, University Groningen, The Netherlands

[64] Willems, J.C. (1979) : System theoretic models for the analysis of physical systems, *Ricerche di Automatica*, Vol. 10, pp. 71-106

[65] Willems, J.C. (1983) : Input-output and state space representations of finite-dimensional linear time-invariant systems. *Lin. Alg. Appl.*, Vol. 50, pp. 581-608

[66] Willems, J.C. (1986) : From time series to linear system. Part I: Finite-dimensional linear time invariant systems, *Automatica*, Vol. 22, pp. 561-580

[67] Willems, J.C. (1989) : Models for dynamics, *Dynamics Reported*, Vol. 2, pp.171-269, U. Kirchgraber and H.O. Walther (eds.), Wiley/Teubner

[68] Willems, J.C. (1991) : Paradigms and puzzles in the theory of dynamical systems, *IEEE Trans. Aut. Control*, Vol. AC-36, pp. 259-294

[69] Willems, J.C. (1992) : Feedback in a behavioral setting, *Systems, Models and Feedback: Theory and Applications*, Eds. A. Isidori and T.J. Tarn, Birkhäuser, pp. 179-191, 1992

[70] Wimmer, H.K. (1981) : The structure of non-singular polynomial matrices, *Math. Systems Theory*, Vol. 14, pp. 367-379

[71] Wolovich, W.A. (1974) : *Linear Multivariable Systems*, Springer Verlag, New York

[72] Wonham, W.M. (1979) : *Linear Multivariable Control: A Geometric Approach*, Springer Verlag, New York

[73] Zadeh, L.A. and C.A. Desoer (1963) : *Linear System Theory, the State Space Approach*, McGraw-Hill

Index

Systems & Control: Foundations & Applications

Series Editor
Christopher I. Byrnes
School of Engineering and Applied Science
Washington University
Campus P.O. 1040
One Brookings Drive
St. Louis, MO 63130-4899
U.S.A.

Systems & Control: Foundations & Applications publishes research monographs and advanced graduate texts dealing with areas of current research in all areas of systems and control theory and its applications to a wide variety of scientific disciplines.

We encourage the preparation of manuscripts in TeX, preferably in Plain or AMS TeX—LaTeX is also acceptable—for delivery as camera-ready hard copy which leads to rapid publication, or on a diskette that can interface with laser printers or typesetters.

Proposals should be sent directly to the editor or to: Birkhäuser Boston, 675 Massachusetts Avenue, Cambridge, MA 02139, U.S.A.

Estimation Techniques for Distributed Parameter Systems
H.T. Banks and K. Kunisch

Set-Valued Analysis
Jean-Pierre Aubin and Hélène Frankowska

Weak Convergence Methods and Singularly Perturbed Stochastic Control and Filtering Problems
Harold J. Kushner

Methods of Algebraic Geometry in Control Theory: Part I
Scalar Linear Systems and Affine Algebraic Geometry
Peter Falb

H - Optimal Control and Related Minimax Design Problems
Tamer Başar and Pierre Bernhard

Identification and Stochastic Adaptive Control
Han-Fu Chen and Lei Guo

Viability Theory
Jean-Pierre Aubin

Representation and Control of Infinite Dimensional Systems, Vol. I
A. Bensoussan, G. Da Prato, M.C. Delfour and S.K. Mitter

Disease Dynamics
Alexander Asachenkov, Guri Marchuk, Ronald Mohler, Serge Zuev

Theory of Chattering Control with Applications to Astronautics, Robotics, Economics, and Engineering
Michail I. Zelikin and Vladimir F. Borisov

Modeling, Analysis and Control of Dynamic Elastic Multi-Link Structures
J. E. Lagnese, Günter Leugering, E. J. P. G. Schmidt

First Order Representations of Linear Systems
Margreet Kuijper